Albrecht on Nitrogen

The Albrecht Papers, Volume IX

by William A. Albrecht, Ph.D.
Edited by Charles Walters

Albrecht on Nitrogen

The Albrecht Papers, Volume IX

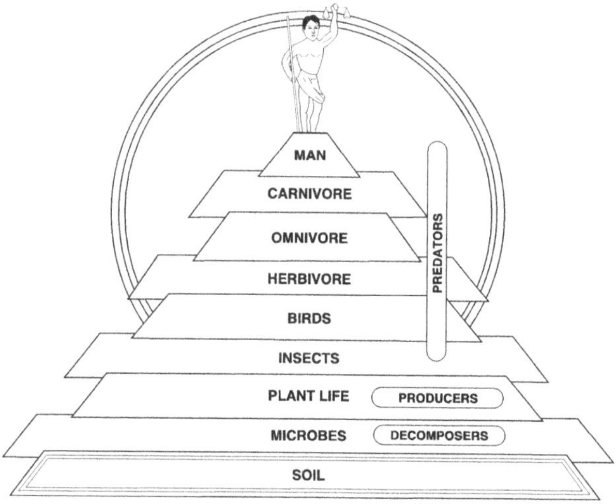

by William A. Albrecht, Ph.D.
Edited by Charles Walters

Acres U.S.A.
Viroqua, Wisconsin

About Acres U.S.A.

Founded in 1971 by Charles Walters, *Acres U.S.A.* magazine emerged from the need to promote ecological farming practices in a time when industrial agriculture was heavily reliant on synthetic fertilizers and pesticides. Inspired by figures like Rachel Carson and Dr. William Albrecht, Walters used the magazine, and later books and conferences, to advocate for sustainable agriculture that prioritized soil health and natural processes. Acres U.S.A. provided a platform for these ideas and helped to popularize alternative methods like cover cropping and integrated livestock management.

Though the agricultural landscape still relies heavily on conventional methods, Acres U.S.A. has been instrumental in the growing movement toward regenerative agriculture. By disseminating knowledge and supporting eco-conscious farmers, the company continues to champion sustainable practices through its publications, conferences, and online resources, contributing to a shift toward a more ecologically sound approach to farming.

Find Out More About Acres U.S.A.

Subscribe to the Online or Print Magazine
https://go.acresusa.com/magazine11365

Attend Our Eco-Ag Conference
https://go.acresusa.com/events11365

Visit the Acres U.S.A. Bookstore
https://go.acresusa.com/bookstore11365

Join the Free Newsletter
https://go.acresusa.com/newsletter11365

A Special Note

The first-rate science of Dr. William A. Albrecht that linked bacterial, fungal and insect attacks to plant malnutrition has been the ongoing lesson that should have earned him a Nobel Prize. The first four volumes in this series amply illustrated that point. Not as well-known are the foundation experiments and studies that led the great professor to his conclusions. Many of these foundation principles—but not all—are contained in this volume suitably titled *Albrecht on Nitrogen*. When the first volume of *The Albrecht Papers* was published, it could not be foreseen that one day the entire corpus of his work would see book-form print. Thus the organization of these papers relied on public response to earlier volumes before the entire project could be pursued. Those who have maximum interest in nitrogen and proteins should seek and find the earlier volumes for added insight.

Albrecht on Nitrogen

Acres U.S.A.
PO Box 351
Viroqua, WI 54665 U.S.A.
970-392-4464
info@acresusa.com • www.acresusa.com

Printed in the United States of America

ISBN 978-0-911311-36-5

Dedicated to the memory of James McHale,
former Commissioner of Agriculture, Commonwealth of Pennsylvania,
a pioneer in the use of eco-friendly nitrogen

About the Author

Dr. William A. Albrecht, the author of these papers, was chairman of the Department of Soils at the University of Missouri College of Agriculture, where he had been a member of the staff for forty-three years. He held four degrees—A.B., B.S. in agriculture, M.S. and Ph.D.—from the University of Illinois. During a vivid and crowded career, he traveled widely and studied soils in the United States, Great Britain, on the European continent, and in Australia.

Born on a farm in central Illinois in an area of highly fertile soil typical of the Cornbelt and educated in his native state, Dr. Albrecht grew up with an intense interest in the soil and all things agricultural. These were approached, however, through the avenues of the basic sciences and liberal arts and not primarily through applied practices and their economics.

Teaching experience after completing the liberal arts course, with some thought of the medical profession, as well as an assistantship in botany, gave an early vision of the interrelationships that enrich the facts acquired in various fields when viewed as part of a master design.

These experiences led him into additional undergraduate and graduate work, encouraged by scholarships and fellowships, until he received his doctor's degree in 1919. In the meantime, he joined the research and teaching staff at the University of Missouri.

Both as a writer and speaker, Dr. Albrecht served tirelessly as an interpreter of scientific truth to inquiring minds and persistently stressed the basic importance of understanding and working with nature by applying the natural method to all farming, crop production, livestock raising, and soil improvement. He always had a specific focus on the effect of soil characteristics upon the mineral composition of plants and the effect of the mineral composition of plants on animal nutrition and subsequent human health.

Dr. Albrecht strove not to be an ivory tower pontificator trying to master and defeat nature, but to be a leader of true science and understand the wondrous ways of nature so we could harness them for the lasting benefit of all. A man of the soil, William A. Albrecht summed up his philosophy as such, "When wildlife demonstrates the soil as the foundation of its health and numbers, is man, the apex of the biotic pyramid, too far removed from the soil to recognize it as the foundation of his health via nutrition?"

Dr. Albrecht was a true student of the characteristics of soil and wasn't timid about his views—be they to a farmer in the field, an industry group or to a congressional subcommittee.

Respected and recognized by scientists and agricultural leaders from around the world, Dr. Albrecht retired in 1959 and passed from the scene in May 1974 as his 86th birthday approached.

About the Editor

Charles Walters was the founder and editor of *Acres U.S.A.*, a magazine he started in 1971 to spread the word about eco-agriculture. A recognized leader in the field of raw materials-based economic research and sustainable food and farming systems, this confirmed maverick saw it as one of his missions to rescue lost knowledge. Perhaps the most important were the papers of Dr. William A. Albrecht, whose low profile obscured decades of brilliant work in soil science. Albrecht's papers, which Walters rescued from the historical dustbin and published in an initial four volumes, continue to provide a rock-solid foundation for the scientific approach to organic farming. Additional volumes of Albrecht's papers were organized and edited by Walters for later publication—the result is shown here with this book.

During his life, Walters penned thousands of articles on the technologies of organic and sustainable agriculture and is the author of more than two dozen books (and co-author of several more), including *Eco-Farm: An Acres U.S.A. Primer, Weeds—Control Without Poisons, A Farmer's Guide to the Bottom Line, Dung Beetles, Mainline Farming for Century 21* and many more.

Charles Walters generously shared his vision, energy and passion through his writing and public speaking for more than 35 years and made it his lifelong mission to save the family farm and give farmers an operating manual that they couldn't live without. *The Albrecht Papers* are an important part of this message.

Charles Walters passed on in January 2009 at the age of 83.

About *The Albrecht Papers*

When the first volume of these papers was issued, no one could foresee the possibility of recovering and publishing all the papers of this great scientist. For this reason the organization of these papers has not followed Dr. Albrecht's work in a calendar sequence, meaning the order of study and investigation. Instead the papers have been organized into topic themes. Here the papers have been grouped to best focus attention and allowed to reciprocate the values upon which all of Albrecht's work rests.

Some papers relating to the general subject considered here might have appeared in earlier volumes, some—likely as not—overlap and appear in later treatments. Nevertheless, an Albrecht paper is valuable regardless of where it turns up. Enjoy and learn.

From the Publisher

A large part of the Acres U.S.A. mission is to preserve and promote the wisdom of those who came before us. Prof. William Albrecht was such a visionary. A collection of his papers was the first publishing entry toward this mission, that book being *The Albrecht Papers*. Charles Walters, our founder and longtime editor, sought out Dr. Albrecht not knowing what lessons were there to be learned. The name Albrecht appeared in journals around the world, yet officials at the University of Missouri where the retired professor kept an office discouraged a meeting citing his age, poor hearing, and the like. Charles Walters visited anyway. What came from these meetings, which soon grew into weekly sessions, was a mentor/student dialogue. Albrecht, ever the patient pedagogue, dispensed the logic, elegance and simplicity of his agricultural systems to an eager mind. And Charles Walters, the publisher and writer, picked up the charge and spread the timeless wisdom of William Albrecht around the world to a new generation of farmers and agronomists.

It's hard to say whether Dr. Albrecht's work would have found its way to light through the efforts of others or if without the republication of Albrecht's papers they would have remained just that, papers to eventually fall into the dustbin of history.

Four volumes of *The Albrecht Papers* eventually came forth. Each had a style and character of its own and each brought new lessons from the master soil scientist to life, but only the most disciplined students of the soil tended to seek out and study these dense works.

Late in his career, semi-retired and legally blind, Charles Walters undertook the Herculean task of reading and sorting the hundreds of remaining papers and articles in Albrecht's archives. He completed that task a few years before his death in 2009. What was left to accomplish was the monumental task of converting damaged and faded copies of articles, some 80 years old, into formats compatible with modern publishing and readable by all.

From file cabinets full of faded photocopies came forth several new volumes of *The Albrecht Papers*, each with a specific focus and theme. In your hands is the beautiful collaboration of the visionary research of Professor William Albrecht and the deft editorial eye of Charles Walters. We hope you enjoy this new creation of William Albrecht and Charles Walters.

Contents

Foreword

That nitrogen equals protein has become a textbook maxim over the last 50 to 75 years largely because texts serve up a conversion formula. This is a handy, albeit unsatisfactory, solution both to fertilization and protein consideration. Of all the Albrecht papers, those concerning nitrogen command the most space.

There is a reason for this. What farmers simply call "N" is both the stuff of life and a regulator of the utilization of potassium, phosphorus, and a raft of other plant food elements. The opening paper presented here deals with fertilization. As the story unwinds, we find the Albrecht key to plant protection and finally protein construction, it being understood that utilization answers the question posed by the character of the nitrogen cycle.

Farmers usually think of nitrogen in terms of shortage, stunted growth, root systems inadequate for support of yield, and leaves yellow or yellowish green, often falling off before their time.

Nitrogen is an outlaw among plant nutrients. It does not reside in the soil, as do so many other plant nutrients, but hovers overhead as unavailable as a drink of water to a thirsty ocean sailor. It has been computed that the overhead nitrogen towering above the land comprises more than 50,000 tons above every acre of land. The papers that follow tell how this resource has to become fixed by soil life before plants can be invited to use it. Fixation takes place when nitrogen becomes a consistent part of amino acid compounds, the foundation for cellular life.

The role of nitrogen—ergo proteins—is examined as under a microscope, sometimes in lay terms, sometimes as scientific fare suitable for the lecture hall. No matter. Readers patient enough to mine Albrecht's thought and proof will encounter the anatomy of the cell, which is fully explained in *Eco-Farm, An Acres U.S.A. Primer*.

It is true, some nitrogen arrives from the heavens via the agency of rainfall, possibly five percent. More is gifted to the farmer by his unpaid work force—bacteria and several organisms. These organisms that fix atmospheric nitrogen in their cells are both symbiotic and non-symbiotic, the first being associated with legumes, the character of which commanded so much attention in earlier volumes, especially in the volume on pastures. The amount of nitrogen fixed by legumes each year staggers the imagination, a number Albrecht and associates computed to be over two million tons.

Non-symbiotic bacteria furnish the rest of nature's nitrogen. There are two concepts of non-symbiotic bacteria. These categories are based on their oxygen requirements, namely the aerobic and the anaerobic. The first is generally represented by the azotobacters, usually operating above pH 6.0 and fixing up to fifty pounds of N per acre. A second type is Clostridium, always residing in the lower pH ranges, always anaerobes. Albrecht and associates computed fixation by this type to total some twenty pounds per acre, usually pinch-hitting for deficient Azotobacter at less than full performance.

Citation of these life forms does not cancel out the contributions made by animal manures, nor does it ignore the nitrogen forms loaded onto the soil from fertilizer factories as a contribution to the industrial model of farming.

Albrecht had a great deal to say about the business of turning a biological procedure into an industrial procedure, such comments being a standing reflection as his investigations caused nature to reveal herself.

The requirement for organic matter in the soil gave the eco-farming movement its first name—organic agriculture. And the suggestion is made that without the proper level of organic matter, farming organically is quite impossible. Again, here is a computation handed off by Albrecht when this editor posed the question. A four percent organic matter soil will hold in storage eight thousand pounds of nitrogen per acre.

Fixation, storage, nitrification—all are covered with clarity and depth, until a transition is made from nitrogen to protein, it being understood that crude protein is a long way from being protein.

The detective story that can be read between ammonification and nitrification—soil conditions, soil life, organic matter, and humus included—is as exciting as any Sherlock Holmes novel.

Withal, a new set of characters fleshes out the pages of this category of Albrecht papers. Albrecht didn't try to corner concepts into airtight compartments. He relied on interplay, wherefore an understanding of nitrogen required an understanding of cation and anion balances, first person familiarity with living residents in the soil, physical and chemical roles of all the actors—nitrosomonas and nitrosococcus, for instance.

The ample supply of Albrecht papers contained in this volume speak for themselves. Therefore, let them speak now.

– Charles Walters, Editor

Fertilizing Soils with Nitrogen[1]

Just as wild animals seem to know how to protect themselves and to be healthy, so plants, in the wild, seem to grow their own protection even for the seed carried over the winter. That the soil fertility should help grow a protection into the seed against grain insects is scarcely surmised until we see it demonstrated. During the season of 1954 some trials with high-nitrogen fertilizers on corn in South Dakota furnished the samples for some careful observations of the failing protection against the lesser grain borer when the corn was fertilized with nitrogen only.

In the illustration given herewith, the small ear, the middle one of the three, represents the size of the ears and suggests the yield of the corn crop on the soil given nitrogenous fertilizer only. This was the treatment suggested by the soil test which indicated no serious fertility deficiency save that the nitrogen might be improved by soil treatment with this nutrient element as fertilizer. The nitrogen was applied with the corn planting in the spring of 1954. The grain was harvested in September.

The ear on the left represents the yield of corn from the plot given, not only nitrogen at the rate used on the other plot, but also this combined with fertilization by phosphates, even though the soil test had not suggested that there was a deficiency in the phosphorus. Both of these grains were grown from hybrid seeds.

The samples from these two soil treatments had been stored for two years when the photograph was made in late September 1956, to illustrate the wide difference in the damages to the seeds by the lesser grain

[1] *Unpublished manuscript, no date.*

Fertilizing a good soil with nitrogen only grew the corn crop represented by the hybrid ear-size of the middle one. In storage of it in contact with the larger hybrid ear on the left—representing the crop on the same soil fertilized with phosphorus as well as nitrogen—the lesser grain borer destroyed the middle ear but bored into the one on the left only at points of contact during storage of two years. The open-pollinated ear on the right grown on fertile soil had one hole after contact for six months.

borer. The ear grown on the soil fertilized with nitrogen only was almost completely riddled with accumulated, floury waste removed two times in advance of the last accumulations exhibited in the illustration. The ear on the left, grown on soil fertilized with phosphates as well as with nitrogen, had been in contact with the badly infested one but was damaged by the grain borer only at the points of contact with the smaller ear, and on the lower side at a few points of contact, apparently, with the shelf. *The ear on the right was grown on well fertilized, highly organic soil from open pollinated seed and had been in contact with the badly damaged ear for six fall and winter months before all were photographed. It showed but one or two grains attacked by the borer.*

It has been previously demonstrated in experimental trials that young soybean plants protected themselves against a fungus attack (suggesting "damping off") as the increasing amount of clay in the soil supplied them with increasing amounts of exchangeable calcium while all other fertility supplies were constant. It has also been demonstrated that the attacks by the leaf-eating thrips insects were prohibited when the nitrogen applied to the soil was in larger amounts; and when the lower amounts of that fertilizer allowed the insect attacks, even those were lessened by increasing amounts of calcium used as fertilizer. This demonstrated also that the fertilizer treatment of nitrogen was better protection against leaf eating insects according as the application was heavier and in combination with calcium. In the case of the stored grain, there was the better protection against the grain borer according as the nitrogen—synonymous with protein—was balanced by some phosphorus, but was inviting destruction almost completely when it was used alone as the fertilizer.

It is significant to remind ourselves that when forage crops use extra nitrogen from the air to build themselves richer in protein—which is the protection or immunity in most living forms—they do so only after all other essentials of fertility are present in the soil in generous amounts. In the humid area, "legumes are hard to grow" or the soils must be well balanced in their fertility if these crops are to fix nitrogen and are to be free of diseases. We dare not rush in to apply nitrogen, then, on the non-legumes—if this reasoning by simile is sound—without anticipating the possible imbalance to be damaging rather than beneficial. If nature fertilizes with nitrogen via the plants' combination of it in the form of organic matter, perhaps the composting of fertilizer nitrogen will be wiser use than its direct application to a crop as salts. We may well ponder that question in relation to what nutritional and protective values the crop finds in the nitrogen treatments as generously applied salts.

CHAPTER 1

Nitrogen and Organic Matter Turnover[1]

In studying the problem of maintaining the organic matter of the soil, the Missouri Agricultural Experiment Station tested a series of cropping systems to discover to what extent the nitrogen and carbon supplies of the soil were either built up or torn down under each. It is the carbon and nitrogen, or the combustible portion, that distinguishes the organic part of the soil from the inorganic, or ash. Of course, there is some ash in all that grows. Thereby all plants put some inorganic matter into their organic residues, and into the humus resulting therefore by decay of them in the soil.

I

This research on soil organic matter, extending over fifteen years, suggested that we cannot expect to build the organic matter of the soil to a high level. We do well to maintain it under cultivation at even a modest level (4). That level is controlled by the climatic setting of the location. That holds true even for soils under no cultivation. For that reason good soil management aims to provide what is called a regular "nitrogen-turnover," or an "organic matter turnover," by putting some crop residue into the soil as regularly as the best soil management in the best farm plan will afford. This carbon and nitrogen of atmospheric origin, buried under the highly mineral soil, represent energy-containing substances for microbial combustion of some—and creation of other—organic compounds. They make the difference between a "living" and a "dead soil."

[1] *Wm. A. Albrecht, "Nitrogen and Organic Matter Turnover," Modern Nutrition 21, no. 5 (May 1968): 9–11.*

The "turnover" of organic matter is taking on more significance now that we are learning more about such organic compounds like the antibiotics originating in soil. They are unusual organic substances created by living cells through which they protect themselves from being consumed by other cells or taken into other protein compounds. Research in soil organic matter is already pointing to this combustible but dynamic part of the soil, as the neglected half of plant nutrition. We have forgotten to study soil organic matter as plant nourishment while studying only the inorganic elements and compounds of them in that function for our crops. We have emphasized the "ash" contents, but have forgotten the organic compounds contained in the crop resulting from the soil's contribution of that "ash." We have not realized that the soil was fertile in terms of both the inorganic and the organic essentials for the growth of the plants.

Only recently has it been demonstrated that the inorganic soil fertility elements serve in plant nutrition more effectively when combined chemically with organic compounds than when used as simple salts. Nature seems to use this procedure of combining the metallic, or inorganic, with the organic to lessen the "salt effect" on the plants of the former, as it were. This organic "tool" seems to help the plant to take out of the soil some inorganic nutrient elements which the root would not take so well otherwise. Metallo-organic compounds have now been chemically synthesized and listed which will "hide away" some inorganic elements within their highly complex organic structure (3). By that they make those inorganic essentials become more effective for serving as plant nourishment. This putting of the simple inorganic fertility element into the center of a large organic molecule, to result in higher fertilizer values thereby, has taken the trade name of "Chelation."

This chelation process makes the feeding of iron, for example, via a spray on the leaves, a very effective cure for the plant's iron shortage or "chlorosis." Chelation is a means, also, of using the trace elements, manganese, copper, and zinc as plant treatments. It is a help in more accurate control of the amount of trace element used when this is but a small part of a much larger molecule or unit. It is an accurate way of diluting, and thereby giving more nearly correct measure, of the amount applied. It lessens the dangers of using excesses. It is a case of the organic matter being the "shock absorber" for applied inorganic salts.

II

More significant than the technical manipulation of the special chelating compounds now being manufactured, is the recognition of the fact that the soil organic matter has been a natural chelating agent for better plant nutrition during all these past years without our recognition of that special physiological service by this small combustible fraction of the soil body. The decay of organic matter has always been giving organic acids like acetic, lactic, citric, malic, tartaric, and others. Organic decay may be viewed as the production and delivery of these organic agencies for chelating the inorganic elements. On combining with the metallic nutrient elements, like calcium, magnesium, potassium, iron, manganese, copper, zinc and others, they make these inorganic essentials function in plant nutrition with less interference from each other in entering the plant and in functioning there. The *organics* "balance" the combination of the *inorganics* as it were. This balancing via this "natural" chelation serves the soil fertility to the plant better than any prescription writing by nutritionists for the crops. We use the same principle when we put calcium gluconate into the bloodstream with benefit but might do so with damage by using calcium chloride.

Of course, any chelating effect on the inorganic fertilizers we may get from plowing organic matter into the soil with them is only a temporary or transitory result. Microbial activity soon decays the acetic, lactic, gluconic, or other parts of the chelated combination as we well know, for example, when vinegar loses its acetic acidity to be like water. We need, then, to keep putting organic matter into the soil regularly if we are to get these beneficial effects in "mobilizing" the inorganic fertility into organic, chelated combinations for better balance of plant nutrition. We need to maintain a continuous organic matter "turnover."

Recent trials of several specific organic compounds fed to seeds grown into young plants under sterile conditions showed that the plants take up organic substances promptly (5). They use them in the form of compounds rather than as elements. Some of the dry-product compounds of digestion excreted by animals are taken up by plants. They suggest themselves as "starter" compounds in the plant's production of its own hormones; in its formation of some commonly deficient parts of protein; and, possibly, in its synthesis of other organic essentials for the growth of the plant (1).

Not so recently an experiment was reported for sunflower plants of which the root system of each plant was divided between two containers.

The inorganic nutrients put into one were taken up and used most efficiently by the plant when some humus extracted from soil was put into the other container (2). This fact gives the suggestion that the plant has its own tools for chelating some inorganic elements by the root system as a whole. That is possible provided the organic compounds for that chelating action are found somewhere in the soil by at least some few roots. The soil organic matter acts as a kind of container within which the soil's inorganic fertility may be gathered and shipped from the soil into the respective place of its function within the plant. Here again is the suggestion that the organic matter of the soil is a decided help in "balanced" nutrition—both organic and inorganic—of the crops we grow.

That the organic matter of the soil in its chelating service is more than a kind of "shipping crate" for the ash elements, is a new fact which will probably emphasize itself more as we study crop quality in relation to soil organic matter, especially in relation to chelated fertilizers. Just what these chelater compounds make within the plant, or just what happens to their by-products, is still unknown. But when potatoes turn black on cooking, supposedly because some element remains in simple inorganic form through failure to be organically combined into another unit, we may appreciate deficiencies of the organic soil fertility as well as that of the inorganics of common fertilizers. The food quality of what we grow may be much more closely connected with the organic matter of the soil than we are now ready to believe. That may be especially true for the quality of the proteins in terms of the complete array of the amino acids required for animal and human nutrition.

While we have been cultivating our soils so intensively and "burning out" their organic matter so rapidly, we may have been depleting a very important safety factor for quality as well as quantity of production, namely, the organic part of nutrition required by plants. Research in soil organic matter is now studying the effects on soil microbes and on crops by the chelated or the metallo-organic compounds, to say nothing of those by the organic compounds alone. These are the neglected part of plant nutrition research emphasizing their significance much as trace elements were coming in for some ten or so years ago. There is the suggestion that careful research on the importance of the soil organic matter as a fertility factor in plant nutrition will do much in giving higher quality in what we grow as feed and food for ourselves, to say nothing of higher quantity.

References Cited

1. Albrecht, W. A. Protein Deficiencies Via Soil Deficiencies. II. Experimental Evidence. Oral Surgery, Oral Medicine, and Oral Pathology 5: 483–99, 1952.
2. deKock, P. C. Influence of Humic Acids on Plant Growth. Science 121: 473–74, 1955.
3. Haertle, E. J. New Horizons in the Application of Chelation to Agriculture. Down to Earth (Dow Chemical Company) II (No. 1) 6–9, 1955.
4. Miller, M. F. Studies in Soil Nitrogen and Organic Matter Maintenance. Missouri Agr. Expt. Sta. Res. Bul. 409: 1–32, 1947.
5. Sheldon, V. L. The Protein Role of Nitrogen in Organic Matter. Progress Report Missouri Agr. Expt. Sta., 1955.

CHAPTER 2

Vitamin C and Nitrogen Fertilizers[1]

The effects of soil fertility on the nutrient or dietary value of vegetable crops deserve increasing attention in experimental studies. Many investigators are agreed that vegetables are nutritionally superior when produced on soils well supplied with the essential fertility elements, and nutritionally inferior when grown on infertile soils.

With increased plant yields due to the applications of fertilizers, there is usually associated an increase in the total content and concentration of minerals in the plants, but not so frequently of vitamins. If the functions of vitamins in the plant are those of catalytic agents, as described by Schopfer and others (32 pp. 194–197), one would not necessarily expect the concentration of these catalysts to be greater, merely because of an increase in plant size. Rather, a rise in the concentration of the catalyst might indicate a natural plant mechanism to make more effective some element that is deficient in the soil, much as the thyroid enlarges in the human body in an attempt to produce more thyroxine when iodine is deficient, and as parathyroid glands increase their hormone production under calcium deficiencies (6).

This report, which confirms the above hypothesis, gives evidence of an inverse relationship between the concentration of vitamin C in plant tissue and nitrogen supplied as fertilizer, together with the usually positive correlation between this soil treatment and the yields of spinach and Swiss chard. In short, the experiments indicate that the concentration of vitamin C

[1] *S. H. Wittwer, R. A. Schroeder, and Wm. A. Albrecht, "Vegetable Crops in Relation to Soil Fertility: II. Vitamin C and Nitrogen Fertilizers," Journal Series no. 967 (Missouri Agricultural Experiment Station, November 6, 1944).*

in these leafy green vegetables increases as the fertility of the soil with respect to nitrogen decreases.[2]

Historical

Numerous studies have been reported wherein applications of different fertilizer caused increase in the concentration of vitamin C (13, 14, 15, 16, 27, 28), whereas other reports concluded that they do not bring about such increases (19, 23, 31, 34). And now considerable experimental work indicates that the application of fertilizers, particularly nitrogenous ones, may reduce rather than increase the concentration of vitamin C in many vegetable and fruit crops.

Some of the earliest work performed was that of Bracewell, Hoyle, Wallace, and Zilva (4, 5, 35) on the antiscorbutic potency of apples. They concluded that the vitamin was in greater concentration in apples of low nitrogen supply. With all varieties tested, under the various conditions of nitrogen fertilization, an inverse relationship existed between the nitrogen content of the fruit and the concentration of vitamin C. Similar conclusions were reached by Kessler (20), who reported that fertilizing apple trees with nitrogen depressed the ascorbic acid content of the fruit progressively as the amount of fertilizer was increased. Hahn and Görbing (9) obtained comparable results with spinach.

In several investigations on tomatoes the influence of fertilizers on vitamin content could be noted. Ott (26) reported that tomatoes grown without fertilizer contained more vitamin C than those fertilized with nitrogen, phosphorus, and potassium. Similarly, Maclinn and Fellers (24) found that tomatoes grown under a low nitrogen supply contained 96–130 units of vitamin C, whereas those under high nitrogen had 74–114 units, but the effects of light and nitrogen were not separated in this experiment. More recently very extensive experiments on the tomato have been conducted by Lyon and his co-workers (11, 22). In their nutritional studies, plants grown with sulfate and nitrogen deficiencies resulted in fruit of a higher vitamin C content. The lack of iron produced a significant rise in ascorbic acid.

---

[2] *In this study, concentrations of vitamin C in the leaves and stems and not in the fruit or seed are considered, because vegetative parts portray a truer index of the influence of variable soil fertility than do reproductive organs, which are necessarily more constant in their chemical composition.*

Again verifying this inverse relationship, Hamdallah (10) found that the vitamin C content of plants which were grown without iron or magnesium was as great as or greater than that of plants fed normally, even though the former were smaller and chlorotic in appearance. In a study of Sudan grass, Wynd (36) reported that the total yield of vitamin C per acre varied inversely as the yield of dry matter and the delivery of nitrogen by the crop. Reder, Ascham, and Eheart (29) studied the effects of fertilizers and environment on the ascorbic acid content of turnip greens and found that potassium and nitrogen produced significant decreases in ascorbic acid. A recent report on grapefruit by workers (18) in Arizona demonstrated an inverse relationship between vitamin C in the juice and nitrogen fertilization.

A careful examination of the literature reveals that environmental factors other than soil nutrient deficiencies may influence the ascorbic acid content. Climatic or other environmental conditions unfavorable for growth frequently result in increased concentrations of vitamin C. Sunny and dry weather, which may reduce the rate of vegetative development, is more conducive to the accumulation of vitamin C than are cloudy and rainy periods (29). Strong light under some conditions has long been recognized as an inhibitor of plant growth, yet light is very influential in increasing the vitamin C content in the plant (12, 30). It is reported (2, 21) that crops grown at high altitudes have an ascorbic acid content sometimes two to three times that of those at lower elevations. Here, again, the high vitamin content is associated with stunted plants grown in an environment unfavorable for maximum vegetative development.

Ascorbic acid is of universal occurrence in growing plants, and it has been demonstrated that there is an enormous increase in its concentration within the germinating seed (25, 30). Bonner and Bonner (3) and Dennison (7) have suggested that it is an essential factor in plant growth. Accordingly, its presence in the plant has been considered something other than merely an accumulation of one of the by-products of metabolism, since it has also been suggested that vitamin C serves as a catalyst in respiration (17, 33). As such, it is allied closely to some of the mineral elements (Fe, Cu, Mn); both are similarly concerned with oxidative enzyme systems. This relationship has been demonstrated, in part, by Elliott and Libet (8), who have shown that an ascorbic acid–iron system exists in the oxidation of phospholipids. Ascorbic acid and a very small amount of iron when used together caused great stimulation of respiration; either one used alone produced some acceleration.

Procedure and Methods

Spinach, the Bloomsdale Long Standing variety, was grown under twenty levels of soil fertility in the greenhouse during the winter of 1941–42. Variable amounts of exchangeable ions placed on the clay subsoil of Putnam silt loam offered a means of controlling the fertility of the soil. Since nearly half of the exchange capacity of this natural clay is taken by hydrogen, various nutrients adsorbed on the clay in exchangeable form may be provided for plants in any desired ratios and quantities, by replacing, more or less completely, the hydrogen by the selected cations and by using the proper amount of the prepared clay in its admixture with sand. The stability of the clay and its naturally high hydrogen content make its use, by simple additions of cations as exchanges for its hydrogen, very convenient.

A series of clay aliquots was prepared by adding calcium acetate to provide 0, 5, 10, 20, and 40 m.e. of calcium. Then from four aliquots of each of these levels of calcium was prepared a nitrogen series by adding 5, 10, 20, and 40 m.e. of nitrogen as ammonium nitrate. This provided, then, twenty different soil treatments giving four levels of nitrogen, each of which had five variable amounts of calcium combined with it as additions to the supply native in the initial clay. To each of these individual treatments were added other nutrients in constant quantities. These additions consisted of 20 m.e. each of potassium and phosphorus and 6 m.e. each of magnesium and sulfate.

The quantity of subsoil clay required to provide the exact exchange capacity for the added nutrients in each treatment was determined beforehand in terms of the known qualities of the clay. This Putnam subsoil material was then mixed under moisture with the particular nutrients and homogeneously blended with pure white quartz sand. Ten replicates of each treatment, or a total of 200 of the mixtures of sand and clay, each in a 1-gallon glazed crock, were prepared. One plant was grown in each. The treatments were randomized on benches and the crocks spaced sufficiently to eliminate light variations. Five of the replicates served for determination of the concentration of vitamin C, while the other five were used for mineral analyses. Entire plants with the exception of roots were utilized.

In the vitamin C determinations, sampling was adjusted such that one plant from each of the twenty treatments was collected at a time, the analyses being completed within 2 to 3 hours after dismantling. This procedure was repeated on consecutive days until the five plants in each treatment were analyzed. In preparation for analyses the material was ground homogeneously in a Waring blender, and vitamin C was determined by modifying slightly the dye reduction method of Bessey and King (1). The vitamin concentration was expressed in milligrams of ascorbic acid per 100 gm. of fresh material.

For the mineral analyses, the plants were collected along with those taken for determination of vitamin C and prepared according to usual methods of harvesting, washing, drying, and preservation. Both fresh and dry weights were recorded. Some determinations were made spectrographically, and others were by regular methods recognized as standard laboratory procedures.

Results

The outstanding features of the crop behavior as influenced by the different fertility levels of the soil were the negative correlation of the concentration of vitamin C with the yields (fresh weights) of the crop; the yields of the spinach as a more direct function of the amount of applied nitrogen; the more decided response by the crop to the different levels of nitrogen than to those of calcium; and the relations in concentrations of calcium and phosphorus with that of the vitamin C.

TABLE 1: Crop yield, and concentration of ascorbic acid and nutrient elements of spinach under variable levels of calcium and nitrogen offered in the soil*

Treatment number	Applied		Ascorbic acid/100 gm.†	Yield/10 plants	Mineral analyses of plants					
	N	Ca			N	Ca	P	K	Mg	Mn
	m.e.	*m.e.*	*mgm.*	*gm.*	*per cent*	*per cent*	*per cent*	*per cent*	*per cent*	*per cent*
1	40	40	70.5 ± 4.7	234.35	6.20	0.71	0.64	8.19	1.25	0.018
5	40	20	84.2 ± 3.6	320.49	6.30	0.69	0.77	7.47	1.03	0.025
9	40	10	88.6 ± 8.5	179.05	6.20	0.64	0.52	7.63	1.05	0.038
13	40	5	82.4 ± 4.8	128.55	6.20	0.79	0.61	7.80	1.07	0.025
17	40	0	120.6 ± 11.1	67.75	6.15	0.63	0.72	7.41	1.06	0.028
Average.......			89.3 ± 5.0	186.04	6.21	0.69	0.65	7.70	1.09	0.027

2	20	40	95.1 ± 2.8	170.80	4.90	0.80	0.82	8.29	0.99	0.020
6	20	20	108.1 ± 8.1	220.05	5.20	0.64	1.06	8.02	0.85	0.034
10	20	10	87.8 ± 8.1	229.00	5.20	0.64	0.76	6.86	0.91	0.031
14	20	5	81.2 ± 7.0	205.10	5.40	0.59	0.72	7.49	0.83	0.034
18	20	0	124.5 ± 3.8	197.32	5.40	0.50	0.77	7.88	0.92	0.033
Average........			99.4 ± 4.6	204.45	5.22	0.63	0.83	7.71	0.90	0.030
3	10	40	170.2 ± 13.6	85.03	3.20	1.13	1.91	7.92	0.68	0.032
7	10	20	134.1 ± 10.1	108.45	3.10	0.75	1.41	8.39	0.70	0.034
11	10	10	113.4 ± 11.7	138.50	3.75	0.66	1.05	7.39	0.83	0.046
15	10	5	184.4 ± 8.5	153.09	4.95	0.68	0.88	8.07	0.89	0.033
19	10	0	126.5 ± 12.9	99.90	3.85	0.45	1.01	6.92	0.53	0.030
Average.........			145.7 ± 8.1‡	116.99	3.77	0.73	1.25	7.74	0.73	0.035
4	5	40	159.9 ± 8.2	36.49	4.05	1.23	2.61	8.31	0.15	0.016
8	5	20	181.9 ± 18.7	65.37	3.35	0.86	1.96	8.68	0.87	0.022
12	5	10	144.0 ± 19.5	114.90	4.60	0.86	1.08	8.23	0.93	0.034
16	5	5	145.6 ± 10.5	60.10	3.15	0.62	1.31	6.71	0.54	0.049
20	5	0	178.4 ± 11.0	76.70	4.20	0.59	1.26	6.86	0.62	0.040
Average...........			162.0 ± 7.6‡	70.71	3.87	0.83	1.64	7.76	0.62	0.032

*Ascorbic acid per 100 gm. fresh weight, yield in fresh weight per 10 plants, and mineral analyses in percentage of dry weight.

† Each value represents mean concentration with its standard error of five plants or replicates in each treatment.

‡ Mean concentration is greater beyond 1 per cent level of significance than means of treatments receiving 40 or 20 m.e. of nitrogen.

These facts are evident from the summary of the results with spinach as presented in Table 1, wherein the treatments for spinach are arranged in groups of different calcium levels with the same quantity of nitrogen, and then the values averaged for each nitrogen level. Included with the data for concentrations of ascorbic acid and yields of plants are the analyses for six soil fertility elements, viz., nitrogen, calcium, phosphorus, potassium, magnesium, and manganese. It is possible to rearrange these data into groups of four different nitrogen levels, averaged for each constant calcium group, to show the influences of the variable calcium. By this means, its effect regardless of nitrogen can be measured.

That the vitamin concentration in the spinach was increased as the nitrogen supplied in the soil was decreased is evident from the table. The yields suggest the characteristic sigmoid curve of decided increase at certain increments in the levels of offered nitrogen, and then no further increase—possibly decrease—at the higher levels of this applied nutrient. The corresponding yields expressed on a dry weight basis portrayed the same correlations with nitrogen additions and vitamin C concentrations as did the yields in terms of fresh weights, and accordingly are not given.

TABLE 2: Ascorbic acid and calcium in spinach crop as related to yield and nitrogen applied

Nitrogen applied ....m.e.	40	20	10	5
Yield/10 plants........................gm.	186.0	204.5	117.0	70.7
Ascorbic acid/10 plants........mgm.	166.1	203.2	170.5	114.6
Calcium/10 plants................mgm.	1412.0	1416.8	939.4	645.6

Striking correlations of vitamin C with the mineral composition are evident. When the concentrations of ascorbic acid were high in spinach, those of nitrogen and magnesium were low. Neither potassium nor manganese showed any connection with vitamin C. Of added interest is the fact that the concentrations of calcium and phosphorus—usually associated more prominently with anabolic than with respiratory processes—are parallel with those of ascorbic acid.

Some thought may well be given to the fact that the supplies of calcium and phosphorus were constant for each decreasing nitrogen level in the soil, yet those two nutrients moved into the crop in amounts that ran almost parallel with the quantities of vitamin C synthesized. Also of interest is the fact that the total vitamin C accumulation in the crop was less at 40 than at 10 m.e. of applied nitrogen, yet yields were increased by 59 per cent at the higher nutrient level (table 2). Total accumulations of ascorbic acid and calcium suggest that as deficiencies in the soil fertility (nitrogen) produced lower yields of crops, the plants were richer not only in their concentration of this vitamin, but also in calcium and phosphorus. The percentages of total nitrogen, of course, were decidedly lower.

This particular chemical behavior seems reasonable in view of other works dealing with vitamin C and mineral nutrition. Frequently among those reported, there were indications that the greater concentrations and yields of vitamin C occurred, not when optimum conditions of fertility were at hand, but under mineral deficiencies.

It is conceivable that with limitations in the mineral supply, the vitamin C content might increase, since, as previously pointed out, ascorbic acid and some of the minor elements (Fe, Cu, and Mn) "function as coenzymes or fragments of coenzymes" (32) in cellular oxidations. Might the increase in the vitamin concentration of plants serve to offset, in part, deficiencies of essential mineral nutrients, and conversely, an optimum mineral supply in the soil depress the vitamin concentration? Plants may be equipped with more than one mechanism of accomplishing a given reaction or completing an essential process for which, if one means fails or becomes limiting, a substitute may be called in.

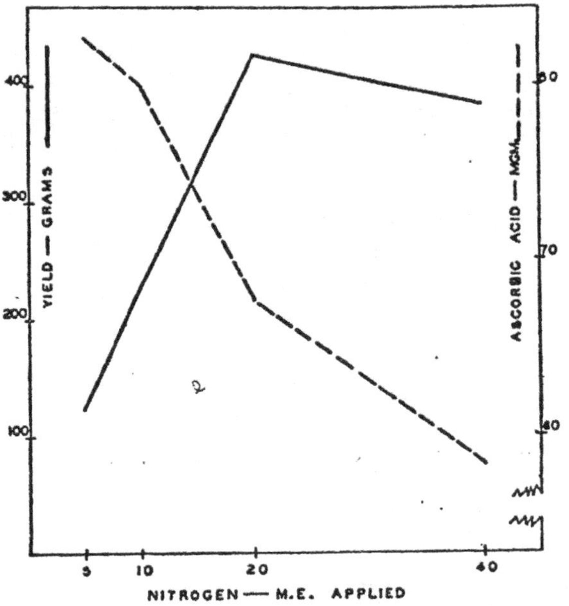

Fig. 1. Concentrations of ascorbic acid and vegetative yields of Swiss chard as related to nitrogen additions to the soil.

As a confirmation of the results obtained for spinach, Swiss chard was grown during the winter of 1943–44. The experimental conditions were identical to those previously used for spinach. Similar clay subsoil and ratios of calcium and nitrogen were utilized. The summarized results of the influence of different levels of soil nitrogen on the yield of the crop and its ascorbic acid composition, as given in Figure 1, portray a repetition of the data already presented for spinach. Again, it was found that with a depression of yields due to deficiencies in soil minerals, the percentage of ascorbic acid in the crop increased; and, conversely, with increased growth and production of vegetative bulk, the concentration of vitamin decreased.

Summary and Conclusions

A careful review of the literature indicates that a high rather than a low vitamin C concentration in plants is associated with a reduction in yield due to nutrient deficiencies, particularly nitrogen. Since certain minerals and ascorbic acid play similar roles as catalysts in plant metabolism, it is suggested that the increase in vitamin C may be a secondary mechanism of the plant to overcome unfavorable conditions of mineral nutrition.

Evidence is presented which indicates that the concentration of vitamin C in leafy green vegetables increases as the fertility of the soil with respect to nitrogen decreases. The following facts were established by the investigation:

There was a decrease in the concentration of ascorbic acid with increasing quantities of nitrogen applied as fertilizer.

A negative correlation was indicated between the yields of the crop and its concentration of ascorbic acid.

There was a decided crop response to nitrogen—as reflected by yields, vitamin C concentration, and mineral content—in some instances quite independent of the calcium levels with which the nitrogen was combined.

The concentrations of calcium and phosphorus within the plant suggest their relation with those for vitamin C.

References

1. Bessey, O. A., and King, C. G. 1933. The distribution of vitamin C in plant and animal tissues, and its determination. *Jour. Biol. Chem.* 103: 687–698.
2. Blagoveshchenskii, A. V. 1937. The influence of conditions of growth on the vitamin C content of certain plants. *Bul. Biol. Med. Exp.* 3: 189–190.
3. Bonner, J., and Bonner, D. 1938. Ascorbic acid and the growth of plant embryos. *Proc. Natl. Acad. Sci.* 24: 70.
4. Bracewell, M. F., Hoyle, E., and Zilva, S.S. 1930. The antiscorbutic potency of apples. *Biochem. Jour.* 24: 82–90.
5. Bracewell, M .F., Wallace, T., and Zilva, S. S. 1931. The antiscorbutic potency of apples. III. *Biochem. Jour.* 25: 144–146.
6. Campbell, I. L., and Turner, C. W. 1942. The relation of the endocrine system to the regulation of calcium metabolism. Missouri Agr. Exp. Sta. Res. Bul. 352.

7. Dennison, R. 1940. Growth response of plants to riboflavin and ascorbic acid. *Science*, 92: 17.
8. Elliott, K. A. C., and Libet, B. 1944. Oxidation of phospholipid catalyzed by iron compounds with ascorbic acid. *Jour. Biol. Chem.* 152: 617–626.
9. Hahn, F. V., and Goerbing, J. 1933. Vitaminstudien: VI. Einfluss der Düngung auf den Vitamingehalt von Spinat. *Ztschr. Untersuch. Lebensmitl.* 65: 601–616.
10. Hamdallah, Abu El Wafa. 1939. Vitamin C—Gehalt Eisen—bzw. Magnesium—frei gezogener Pflanzen. *Protoplasma* 32: 31–43.
11. Hamner, K. C., Lyon, C. B., and Hamner, C. L. 1942. Effect of mineral nutrition on the ascorbic-acid content of the tomato. *Bot. Gaz.* 103: 586–616.
12. Hamner, K. C., and Parks, R. Q. 1944. Effect of light intensity on ascorbic content of turnip greens. *Jour. Amer. Soc. Agron.* 36: 269–273.
13. Hester, J. B. 1940. The influence of soil type and fertilization upon yield and composition of tomatoes. *Soil. Sci. Soc. Amer. Proc.* 5: 281–283.
14. Hester, J. B. 1941. Manganese and vitamin C. *Science*. 93: 401.
15. Lido, J. B. H. 1936. The influence of fertilizers on the carotene and vitamin C content of plants. *Biochem. Jour.* 30: 2307–2312.
16. Isgur, B., and Fellers, C. R. 1937. A preliminary study of the relationship between vitamin C content and increased growth resulting from fertilizer applications. *Jour. Amer. Soc. Agron.* 29: 890–893.
17. James, W. O., and Cragg, J. M. 1943. The ascorbic acid system as an agent in barley respiration. *New Phytol.* 42: 28–44.
18. Jones, W. F., et al. 1944. A note on ascorbic acid: nitrogen relationships in grapefruit. *Science*. 99: 103–104.
19. Karikka, K. J., Dudgeon, L. T., and Hauck, H. M. 1944. Influence of variety, location, fertilizer, and storage on the ascorbic acid content of potatoes grown in New York State. *Jour. Agr. Res.* 68: 49–63.
20. Kessler, W. 1939. Über den Vitamin C-Gehalt deutscher Apfelsorten und seine Abhängigkeit von Herkunft, Lichtgenuss, Düngung, Dichte des Behanges und Lagerung. *Gartenbauwiss.* 13: 619–638.
21. Kostenko, V. D. 1943. Content of vitamin C in cultivated and wild plants growing in high regions of Pamir. *Compt. Rend.* [Doklady] 38: 42–43.

22. Lyon, C. B., Beeson, K. C., and Ellis, G. H. 1943. Effects of micronutrient deficiencies on growth and vitamin content of the tomato. *Bot. Gaz.* 104: 495–514.
23. Mack, G. L., Tapley, W. T., and King, C. G. 1939. Vitamin C in vegetables: X. Snap beans. *Food Res.* 4: 309–316.
24. Maclinn, W. A., and Fellers, C. R. 1938. Ascorbic acid (vitamin C) in tomatoes and tomato products. Mass. Agr. Exp. Sta. Bul. 354.
25. Neubauer, M. 1939. Das Vitamin C in der Pflanze. *Protoplasma* 33: 345–370.
26. Ott, M. 1938. Pflanzenqualität, Volksernährung und Dügung. *Forschungsdienst* 5: 546–552.
27. Pfaff, C., and Pfützer, G. 1937. Über den Einfluss der Ernahrung auf den Carotin-und Ascorbinsaüre-Gehalt verschiedener Gemüse und Futterpflanzen. *Angew. Chem.* 50: 179–184.
28. Potter, M. T., and Overholser, E. L. 1933. The vitamin C content of the winesap apple as influenced by fertilizers. *Jour. Agr. Res.* 46: 367–373.
29. Reder, R., Ascham, L., and Eheart, M. S. 1943. Effect of fertilizer and environment on the ascorbic acid content of turnip greens. *Journ. Agr. Res.* 66: 375–388.
30. Reid, M. E. 1938. The effect of light on the accumulation of ascorbic acid in young cowpea plants. *Amer. Jour. Bot.* 25: 701–711.
31. Scheunert, A., Reschke, J., and Kohlemann, E. 1940. Über den Vitamin C Gehalt der Kartoffeln: IV. Mitteilung: Über den Einfluss verschiedener Düngung. *Biochem. Ztschr.* 305: 1–3.
32. Schopfer, W. H. 1943. Plants and Vitamins. Chronica Botanica Co., Waltham, Mass.
33. Szent-Györgyi, A. V. 1939. On Oxidation, Fermentation, Vitamins, Health and Disease. Williams & Wilkins Co., Baltimore.
34. Wachholder, K., and Nehring, K. 1938. Über den Vitamin C-Gehalt verschiedener Kartoffelsorten und seine Abhängigkeit von Düngung. *Bodenk. u. Pflanzenernähr. (n.s.)* 9–10: 708–724.
35. Wallace, T., and Zilva, S. S. 1933. The antiscorbutic potency of apples: VI. *Biochem. Jour.* 27: 693–698.
36. Wynd, F. L. 1942. Comparison of the efficiency of single applications with repeated top dressing of nitrogenous fertilizers in increasing the yield of dry matter, nitrogen, and vitamin C (ascorbic acid) of Sudan grass. *Plant Physiol.* 17: 645–651.

CHAPTER 3

Oxalate Content and Nitrogen Fertilization[1]

Green leafy vegetables may be one of our richest sources of dietary calcium. Although these crops generally make adequate growth over a wide range of soil differences, the fertility level may presage a wide range in their mineral composition. Of late, the relative availability or digestibility of organo-calcium and other mineral-organic complexes in many of these vegetables has been the subject of numerous investigations, such as those of Tisdall and Drake (1938), Spiers (1939), and Kohman (1939). A summary of this research, which has involved feeding tests conducted on both man and smaller animals, shows rather conclusively that many common leafy vegetables, including spinach, Swiss chard, beet greens, and New Zealand spinach, may be worthless as contributors of calcium and magnesium and in addition, if consumed in moderate quantities, may render unavailable a large percentage of the calcium normally furnished in the diet by other foods. This undesirable feature is attributed to the large amounts of oxalic acid which probably combine not only with calcium in the plant but also with the magnesium to form insoluble oxalates, Pierce and Appleman (1943), making these two essential elements from the soil valueless so far as human nutrition is concerned.

Since these oxalate-containing foods now play a prominent, though perhaps not a desirable, role in the human diet, an analysis of plants for the chemically equivalent relationships between their oxalate contents on the one hand, and their concentrations of calcium and magnesium on the

[1] *S. H. Wittwer, Wm. A. Albrecht, and H. R. Goff. "Vegetable Crops in Relation to Soil Fertility: III. Oxalate Content and Nitrogen Fertilization." Journal Series no. 976 (Missouri Agricultural Experiment Station, March 19, 1945).*

other, was deemed essential. Moreover, the wide fluctuations in oxalic acid known to occur in a given crop of the same variety, as indicated by Kohman (1939), prompted a further inquiry into possible soil factors, causal of this variation, other than those of calcium and pH already outlined by Schroeder and Albrecht (1942). The following investigation represents an analysis and study of the oxalate and mineral contents of Swiss chard (variety Lucullus), which was subjected to 16 levels of soil nitrogen and calcium as combinations of four different amounts of each.

Experimental Procedure

Satisfactory control in providing the desired fertility gradients was achieved by using the subsoil of Putnam silt loam as a carrier for exchangeable soil nutrients. This material, with its high content of colloidal acid clay, has the semblance of natural soil and permits the uptake of nutrients by the plant through contact exchange.

A series of soil treatments was prepared by supplying calcium and nitrogen levels each of 5, 10, 20, and 40 milliequivalents (m.e.) with all possible (16) combinations of two nutrients. Accordingly, for each level of calcium there were four levels of nitrogen, and for each level of nitrogen there were four of calcium. The calcium and nitrogen were provided by adding variable quantities of calcium acetate and ammonium nitrate, respectively, to the correspondingly required clay aliquots, replacing the adsorbed hydrogen thereon with cationic nutrients. Other materials supplied in constant amounts consisted of 20 m.e. each of potassium and phosphorus, and six m.e. each of magnesium and sulfate. The resulting clay samples, almost completely saturated with various ratios of exchangeable calcium and nitrogen, were, by using moisture, blended with pure white quartz sand. Each treatment was replicated 10 times, the plants being grown in one-gallon glazed crocks in which one plant for each container was retained till harvest time. The nutrients and the clay supplied for each plant with the resulting pH are presented (Table 1).

TABLE 1: Nutrients added to clay to provide variable calcium and nitrogen levels								
Treatment	Milliequivalents per plant						Clay per plant gm.	Resulting pH
	Ca	N	P	K	Mg	S		
Ca Group I[1]								
1	40	40	20	20	6	6	716.6	6.5
2	40	20	20	20	6	6	633.3	6.8
3	40	10	20	20	6	6	591.6	6.8
4	40	5	20	20	6	6	570.8	7.0
Ca Group II								
5	20	40	20	20	6	6	550.0	6.4
6	20	20	20	20	6	6	466.6	6.6
7	20	10	20	20	6	6	425.0	6.6
8	20	5	20	20	6	6	404.1	6.8
Ca Group III								
9	10	40	20	20	6	6	466.6	6.4
10	10	20	20	20	6	6	383.3	6.5
11	10	10	20	20	6	6	341.6	6.6
12	10	5	20	20	6	6	320.8	6.6
Ca Group IV								
13	5	40	20	20	6	6	425.0	6.6
14	5	20	20	20	6	6	341.6	6.4
15	5	10	20	20	6	6	300.0	6.5
16	5	5	20	20	6	6	279.1	6.5

1 In a like manner four N groups may be arranged, each containing four levels of Ca

The experiment was conducted under greenhouse conditions during the late fall and early winter 1943–44. All plants were harvested after a growing period of 90 days, at which time they were judged to be in the most desirable eating condition. The tops were cut, weighed, dried at 65°C. (149°F.), and the total dry matter recorded. After being shredded in a Wiley mill and finely ground in a Merker mill, the dried material was suitable for analyses. The chemical determinations consisted of oxalate, total nitrogen, calcium, magnesium, and phosphorus. All analyses were performed in accordance with regular laboratory methods recognized as official or standard procedures.

Results and Discussion

The total yields as fresh weights and the results of the chemical analyses in terms of percent concentrations in the dried material are assembled (Table 2). The treatments are first arranged into calcium groups, of which the four different nitrogen levels are averaged to show the influence of each level of calcium. Secondly, the treatments are regrouped according to the nitrogen applied, and the mean values of the calcium treatments taken so that the average for each nitrogen level could be recorded.

The most striking features evident from the data presented seem to be the following: (1) the magnitude of yield increase was about equal when either nitrogen or calcium was applied in additive quantities; (2) regardless of the calcium supply, the percentage of oxalate in the crop generally increased as more nitrogen was provided; (3) the percentage of total nitrogen in the crop was directly correlated with that of the oxalate and with the nitrogen supplied in the substrate; (4) the percentages of plant calcium were consistently greater as the calcium in the soil was increased, while the magnesium at first varied directly with the increments of both calcium and nitrogen and then showed a decided drop when the highest level of either was reached; and (5) there were marked differences, owing to soil treatment, in the magnesium and phosphorus percentages although the quantities of these nutrients applied to the soil were constant for all groups.

These results are of special significance in that they demonstrate remarkable variability in the important dietary components within the same crop, as a result of growing it on soil media differing in fertility. Differences in fertility not only gave variances in the quantities of the mineral elements moved from the clay-sand mixture into these plants, but it is likely that much of the fluctuation in the concentrations of the oxalate—a synthetic product of the plant's metabolism—may be accounted for by variability in the fertility of the soil.

Oxalate in chemical combination with calcium forms a very insoluble compound. This is likely true also of its union with magnesium. Since some of the leafy vegetable crops have been reported as containing more than sufficient oxalate to precipitate and put the entire supply of these two mineral nutrients into inert forms, so far as ordinary digestive processes are concerned, it seemed essential in these studies to examine to what degree the oxalate was stoichiometrically balanced by the calcium and magnesium in Swiss chard. These relations are best shown by Fig. 1, wherein the probable fate of the total plant oxalate is portrayed.

Several facts are evident. Stoichiometrically, the oxalate content exceeded by several times that of calcium at all fertility levels. In every instance the total oxalate was in excess of the plant calcium and magnesium combined. This left a sizable surplus of soluble oxalates for possible combination with calcium derived from other foods in the diet. It is also to be noted that the quantity of oxalate in excess of the plant calcium alone, or of the summation of the calcium and magnesium, was directly correlated with the nitrogen offered in the soil. This excess was particularly marked when the calcium supply was low. The excess soluble oxalates diminished as the nitrogen supply decreased. This was true regardless of the calcium level.

TABLE 2: Yield, oxalate content, and mineral composition of Swiss chard when grown under variable levels of nitrogen and calcium

Treatment	Variable m.e. applied		Total yields in fresh weight per 10 plants	Concentrations in per cent of dry weight				
	N	Ca		Oxalate	Total nitrogen	Calcium	Magnesium	Phosphorus
				Ca Group I				
1	40	40	634.6	6.63	4.29	1.032	.477	1.33
2	20	40	486.4	5.58	3.16	0.888	.416	1.30
3	10	40	233.3	4.70	2.58	0.886	.534	1.87
4	5	40	179.8	3.93	2.40	0.748	.320	2.67
Average	...	...	383.5	5.21	3.10	0.889	.437	1.79
				CA Group II				
5	40	20	487.3	6.47	4.61	0.784	.734	1.04
6	20	20	693.5	5.57	3.92	0.676	.864	1.10
7	10	20	243.4	4.79	2.63	0.720	.678	1.05
8	5	20	133.8	4.32	2.06	0.776	.542	1.66
Average	...	...	389.5	5.29	3.31	0.739	.705	1.21

Ca Group III								
9	40	10	391.6	5.63	4.57	0.600	.440	1.06
10	20	10	333.8	6.94	3.97	0.776	.720	0.88
11	10	10	215.2	4.38	2.53	0.824	.552	1.41
12	5	10	116.6	3.78	2.43	0.746	.368	2.04
Average	...	...	264.3	5.18	3.38	0.737	.520	1.35
Ca Group IV								
13	40	5	32.0	6.50	4.15	0.488	.380	1.81
14	20	5	212.7	7.34	3.91	0.820	.646	0.85
15	10	5	171.5	5.00	2.77	0.660	.284	1.11
16	5	5	104.6	3.96	2.25	0.552	.390	1.57
Average	...	...	130.2	5.70	3.27	0.630	.425	1.34
N Group I								
1	40	40	634.6	6.63	4.29	1.032	.477	1.33
5	40	20	487.3	6.47	4.61	0.784	.734	1.04
9	40	10	391.6	5.63	4.57	0.600	.440	1.06
13	40	5	32.0	6.50	4.15	0.488	.380	1.81
Average	...	...	386.4	6.31	4.41	0.726	.508	1.31
N Group II								
2	20	40	486.4	5.58	3.16	0.888	.416	1.30
6	20	20	693.5	5.57	3.92	0.676	.864	1.10
10	20	10	333.8	6.94	3.97	0.776	.720	0.88
14	20	5	212.7	7.34	3.91	0.820	.646	0.85
Average	...	...	431.6	6.36	3.74	0.790	.662	1.03
N Group III								
3	10	40	233.3	4.70	2.58	0.886	.534	1.87
7	10	20	243.4	4.79	2.63	0.720	.678	1.05
11	10	10	215.2	4.38	2.53	0.824	.552	1.41
15	10	5	171.5	5.00	2.77	0.660	.284	1.11
Average	...	...	215.9	4.72	2.63	0.773	.512	1.36
N Group IV								
4	5	40	179.8	3.93	2.40	0.748	.320	2.67
8	5	20	133.8	4.32	2.06	0.776	.542	1.66
12	5	10	116.6	3.78	2.43	0.746	.368	2.04
16	5	5	104.6	3.96	2.25	0.552	.390	1.57
Average	...	...	133.7	4.00	2.29	0.706	.405	1.99

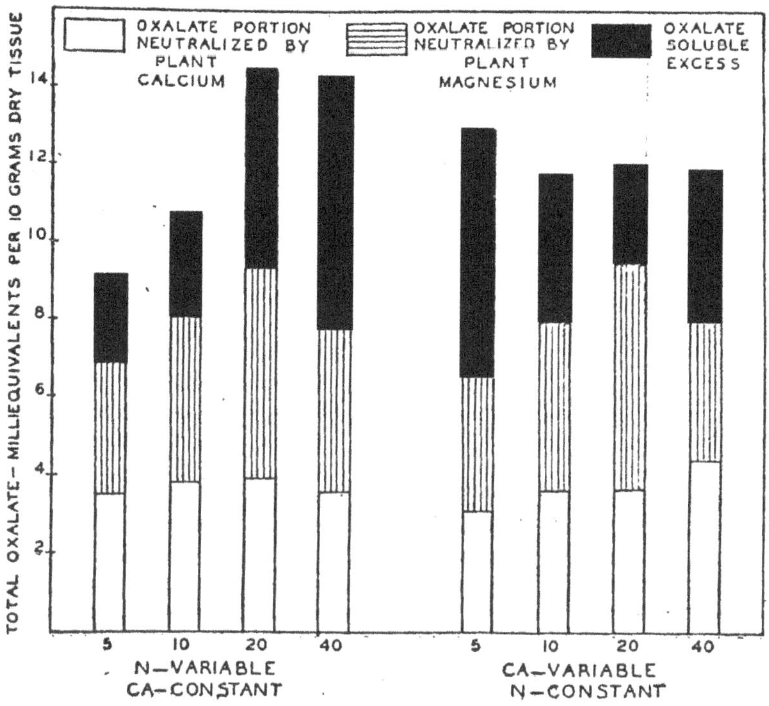

Fig. 1. Probable disposition of oxalate in Swiss chard when grown under variable levels of nitrogen and calcium in the soil.

Though the above relations between plant oxalates and the degree of nitrogen fertilization present a new departure in plant nutrition with respect to the nutritive quality of some leafy vegetables, the basic concept that nitrogen fertilizers may profoundly influence the organic acid composition of other crops was first emphasized by Ruhland and Wetzel (1926). This effect is accentuated if nitrate rather than ammonium nitrogen is used, according to Clark (1936); Wadleigh and Shive (1939); Vickery, Pucher, Wakeman, and Leavenworth (1940); and Blackman and Templeman (1940). Pepkowitz, Gilbert, and Shive (1944) suggest a direct association of oxalic acid formation with nitrate reduction. In the present investigation, ammonium nitrate was used as the nitrogen fertilizer. If nitrate-nitrogen alone had been employed, the results might have proven even more conspicuous.

These data, interpreted in conjunction with those of Schroeder and Albrecht (1942), suggest that soil-fertility practices in the production of Swiss chard and spinach may offer a certain degree of control over the quantities of oxalate produced. One would expect the least disturbance from oxalate when the crops are grown on an acid soil high in exchangeable calcium and low in readily available nitrogen. Wittwer, Schroeder, and Albrecht (1945) have demonstrated that heavy applications of nitrogen as a fertilizer salt seem to lessen their mineral as well as vitamin C contributions in the human diet. With reference to food qualities, it is unfortunate in the production of these vegctables that the yield response is so favorable toward increased nitrogen fertilization, as shown by Schroeder and Wittwer (1944).

Where dietary preferences permit, spinach, Swiss chard, beet greens, and New Zealand spinach may be replaced by other green leafy vegetables known to be free from the harmful effects of oxalate. Such crops include kale, mustard greens, collards, and turnip greens. The initial supply of all basic minerals and vitamins in this latter group of vegetables generally exceeds that of the high oxalate plants, and, containing an insignificant amount of oxalate, they contribute calcium and magnesium in forms that are readily assimilated by the digestive system.

Summary

By a study of the oxalate-calcium and oxalate-magnesium relations in Swiss chard, grown under 16 levels of variable calcium and nitrogen in colloidal clay cultures, it was shown that the concentrations of oxalate in the crops increased as more nitrogen was provided in the substrate. This increase seemed to be independent of the calcium supply.

When the chemically equivalent quantities of oxalate with respect to its possible union with the plant's calcium and magnesium were considered, it was demonstrated that for all fertility levels the total oxalate production was far in excess of its equivalent of the plant's calcium, or even of calcium and magnesium combined. These surpluses of oxalate, beyond precipitation by calcium and magnesium, were progressively of greater magnitude as the nitrogen supply was increased and the calcium decreased.

References

1. Blackman, G. E., and Templeman, W. G., 1940. The intersection of light intensity and nitrogen supply in the growth and metabolism of grasses and clover *(Trifolium repens)*. IV. The relation of light intensity and nitrogen supply to protein metabolism of the leaves of grasses. Ann. Botany 4, 533–587.

2. Clark, H. E., 1936. Effect of ammonium and of nitrate nitrogen on the composition of the tomato plant. Plant Physiol. 11, 5–24.
3. Kohman, E. F., 1939. Oxalic acid in foods and its behavior and fate in the diet. J. Nutrition 18, 233–246.
4. Pepkowitz, L. P., Gilbert, S. G., and Shive, J. W., 1944. The importance of oxygen in the nutrient substrate for plants—organic acid. Soil Sci. 58, 295–303.
5. Pierce, E. C., and Appleman, C. O., 1943. Role of other soluble organic acids in the cation-anion balance in plants. Plant Physiol. 18, 224–238.
6. Ruhland, W., and Wetzel, K., 1926. Zur Physiologie der organischen Säuren in grünen Pflanzen. 1. Wechselbeziehung im Stickstoff—und Säurestoffwechsel von *Begonia semperflorens*. Plants 1, 558–564.
7. Schroeder, R. A., and Albrecht, Wm. A., 1942. Plant nutrition and the hydrogen ion. III. Soil calcium and the oxalate content of spinach. Bull. Torrey Botan. Club 69, 561–568.
8. ———, and Wittwer, S. H., 1944. Vegetable crops in relation to soil fertility. I. Yields of lettuce and spinach as influenced by variable calcium and nitrogen. Proc. Am. Soc. Hort. Sci. 44, 469–472.
9. Spiers, Mary, 1939. The utilization of the calcium in various greens. J. Nutrition 17, 557–564.
10. Tisdall, F. F., and Drake, T. G. H., 1938. The utilization of calcium. J. Nutrition 16, 613–620.
11. Vickery, H. B., Pucher, G. W., Wakeman, A. J., and Leavenworth, C. S., 1940. Chemical investigation of the tobacco plant. VIII. The effect upon the composition of the tobacco plant of the form in which nitrogen is supplied. Connecticut Agr. Expt. Sta. Bull. 442.
12. Wadleigh, C. G., and Shive, J. W., 1939. Organic acid content of corn plants as influenced by pH of substrate and form of nitrogen supplied. Am. J. Botany 26, 244–248.
13. Wittwer, S. H., Schroeder, R. A., and Albrecht, Wm. A., 1945. Vegetable crops in relation to soil fertility. II. Vitamin C and nitrogen fertilizers. Soil Sci. 59, 329–336.

CHAPTER 4

Nitrate Accumulation in Soil as Influenced by Tillage and Straw Mulch[1]

Even though plants may use several forms of nitrogen in their nutrition, the nitrate form still stands out as the most important. The amount of nitrate produced in the soil during the season is more eventual in determining crop production than the total supply of nitrogen. Numerous soil treatments and conditions influence this nitrogen turnover, or this change from the insoluble nitrogen to the soluble form. Among these, the mechanical treatment, or tillage, has been considered significant. The following study was conducted in the hope of getting some information regarding the influence of tillage, such as plowing and surface cultivation, the influence of the straw mulch, and other factors on the nitrate accumulation in the soil.

Soil and Treatments Used

This study deals with one series of 11 plats of a silt loam, classified as a phase of Shelby loam, and 4 plats from a similar series of 10, not far removed, on the same type of soil. The former series was under study for six years, 1920 to 1925, inclusive, and the latter for three years, 1917 to 1919, inclusive. These series included plats with different crops along with different tillage treatments, but only the latter are being considered in this discussion. The treatments and crops on the different plats used are given in Table 1.

[1] *William A. Albrecht, "Nitrate Accumulation in Soil as Influenced by Tillage and Straw Mulch," Journal of the American Society of Agronomy 18, no. 10 (October 1926): 841–53. Paper read as a part of the symposium on "Soil Bacteriology—Nitrification Studies" at the meeting of the Society held in Chicago, Ill., November 17, 1925.*

The soil of the plats is a shallow silt loam about 7 inches deep underlain by a fairly heavy, more or less impervious, silty clay. The surface has a slope of about 5° and is subject to some washing, but receives little added material due to the sod crops above.

The rainfall of the season was 37.6 inches as a mean for the year, with the distribution giving a significant decrease during the summer months suggesting a droughty condition. This is shown in Fig. 3 giving the nitrates in the undisturbed sod soil and the rainfall distribution by months for the 6 years under study and for 37 years. From these plats it was possible to note the general course of nitrate accumulation under an untilled sod in grass and to study the effects on nitrate accumulation by the general operations of seedbed preparation, including spring plowing and various dates of fall plowing for wheat, by surface cultivation with the ordinary corn cultivator, and by the straw mulch in heavier and lighter applications.

TABLE 1: Treatments in study of influence of tillage and straw mulch on nitrate accumulation

Plat No.	Tillage treatment 1919–1925, inclusive	Crop
1	None	Grass sod
2	Plowed July 15th[a]	Wheat continually
3	Plowed August 15th	Wheat continually
4	Plowed Sept. 15th	Wheat continually
5	Spring preparation,[b] mulch, 6 tons straw per acre[c]	No crop
6	Spring preparation, mulch, 2 tons straw per acre	No crop
7	Spring preparation, scraped	No crop
8	Spring preparation, cultivated three times	No crop
9	Spring preparation, cultivated 3 times	Corn
10	Spring preparation, scraped	Corn
11	Scraped	Corn

	1917–1919, inclusive	
3	Spring preparation, mulch, 8 tons straw per acre	No crop
4	Spring preparation, scraped	No crop
5	Scraped	No crop
7	Spring preparation, cultivated after each rain	No crop

[a]When soil plowed cloddy it was disked after next rain making such possible. Plats 2, 3, and 4 were worked before seeding wheat in October.
[b]Spring preparation consisted of plowing in April, occasionally in May, together with harrowing to produce a condition similar to a good seedbed.
[c]The mulch was applied after spring preparation and remained until March of the next year when it was removed to allow soils to dry before plowing.

Methods Used

In the series studied from 1917 to 1919, nitrate determinations were made at intervals of three weeks on the surface 7 inches of soil during the greater part of the three years. On the series during 1921 to 1925 determinations were made every two weeks for almost the entire time, including the winters. Only the surface soil data were taken since data taken on the subsurface showed no significant variation, and since this layer consisting of a silty clay is too impervious to permit much water movement. The samples were taken as 9 to 12 borings per plat of about one seventy-third of an acre area, mixed thoroughly, and triplicate 100-gram samples dried in the oven to determine the moisture. The dried samples were extracted with dilute acid, boiled with alkali to remove the ammonia, and then reduced with De Varda's metal during distillation into standard acid for titration. The nitrates were calculated as pounds of nitrogen per acre of two million pounds of oven-dry soil. The extensive samplings during the six years have given so much data that only the averages for any plat can be given by months or by years. Data of this kind only will be presented.

Experimental: Nitrate Under Sod

The entire area represented by the later series was under cultivation before going into this experiment and sod production was slow the first year. During the remaining time this plat was in a fair sod and was left undisturbed, save for sampling. The nitrate determinations for this sod plat are given as averages by months in Fig. 3.

Fig. 1. Weeds on wheat plats after harvest, as influenced by date of plowing. A, plowed August 15. B, plowed September 15.

Fig 2. Wheat growth as influenced by date of plowing. A, plowed July 15. B, plowed August 15. C, plowed September 15.

The data for sod show clearly the correlation during the growing period of the nitrate accumulation, in this case, with the moisture supply. The curve of nitrates follows the rainfall curve, as averaged for the longer time, better than that for the six-year period under study. The nitrates show an increase with the increased rain and temperature of the spring until the maximum is reached in May. They then decline until August probably through removal by grass growth and decreasing moisture until August, when the increased moisture again gives increased nitrates. During all the time, nitrates are distinctly low under sod, since this crop probably keeps them exhausted to this low point. During the last five years, or after the sod was well established, the maximum nitrate accumulation was 17.9 pounds nitrogen per acre which was reached in September, 1923.

Influence of Spring Plowing

As a means of mixing and aerating soil, the mold board plow is an outstanding implement. For nitrification, which is an oxidative process, plowing should then be a significant operation. That such is the case is shown by the curves in Fig. 4 representing the average monthly nitrate nitrogen in the soil which was unplowed; one which was plowed; and one which was plowed and cultivated. These emphasize the increased nitrate in soil growing corn as a result of aerating the soil through plowing. Similar results were obtained on fallow soils, as shown in Fig. 5 for soils under study three years.

The curves for the unplowed and plowed soils in Fig. 4 show distinct differences in nitrates during the growing season, which averaged 2.16 pounds in favor of the plowed soil from March to August inclusive. The difference for the year as a whole was not so great, however, as was shown by the averages of all the determinations giving 10.29 pounds nitrogen as nitrate for the unplowed soil and 11.95 pounds for the plowed, or a difference for the entire year of about 1.66 pounds. Since the soil samples were taken as far away from the corn plants as possible, it is highly probable that the corn roots reached the point of sampling and removed the nitrates from the plowed soil much earlier in the season than was the case in the unplowed soil. Consequently the data for the latter may be comparably higher which would lessen the difference. This is suggested by the fact that the fallow soils under three years of study, as shown in Fig. 5, give a much wider spread, with an average difference of 6 pounds in favor of the plowed soil, during the months of May to November inclusive.

According to Fig. 4 there is a distinct rise in nitrates in April for the plowed plat while such is not true for the unplowed soil. Since plat 10 was usually plowed during this month one might infer that this operation was responsible for a distinct increase immediately. Such is not the truth, however, since the five-year average for two determinations for a month just after the date of plowing compared with the same average for a month just before the date of plowing is not significantly greater for the plowed than for the unplowed soil. The figures in comparison are 15.02 pounds and 14.44 pounds after and before, respectively, for the plowed soil, and 11.50 pounds and 10.85 pounds for the unplowed soil, giving an increase of 0.58 pound in the plowed and 0.65 pound in the unplowed soil during these two months. This indicates that the operation of plowing on this soil does not give any sudden increase in nitrates, yet serves to bring about conditions giving a general nitrate content in the soil during a long time following, especially through the growing season. Even though this difference becomes almost insignificant during the winter months, the increase during the growing season is large enough to give the plowed soil an annual average nitrate nitrogen content of 11.95 pounds as compared to 10.29 pounds for the unplowed soil.

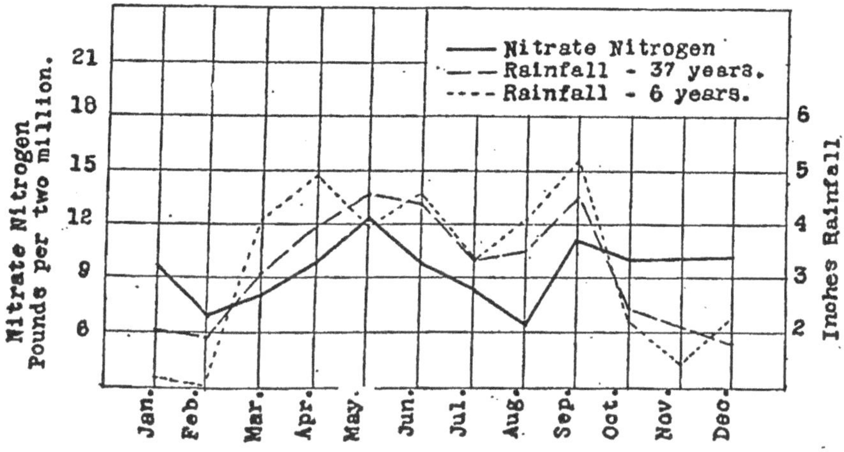

Fig. 3. Nitrates under soil in sod as correlated with distribution of rainfall.

The yields from these plats are in accord with the differences in nitrates, giving distinctly higher yields for plowing than the differences in nitrate would suggest. Yields are given in Table 2.

Influence of Fall Plowing

Fall plowing also shows some interesting results. Three plats in wheat continually were plowed at varying dates, in preparation for this crop, to note the effect of such on the nitrate accumulations. Plat 2 was plowed for wheat on July 15, Plat 3 on August 15, and Plat 4 on September 15. The data are summarized by averages for the months of the year in graphs in Fig. 6. Since the sampling began in May, 1921, and since these plats did not receive their first treatments until the particular month for plowing, the data are considered beginning with the time of the first plowing.

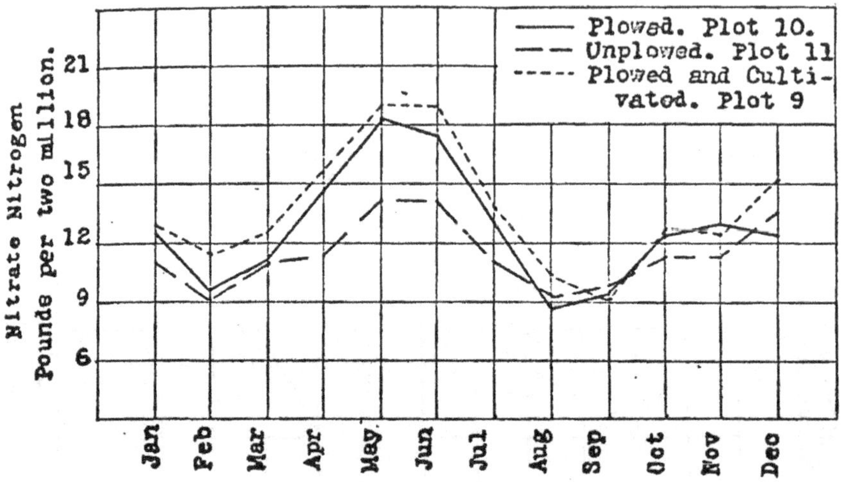

Fig. 4. Nitrates in soil under corn as influenced by plowing.

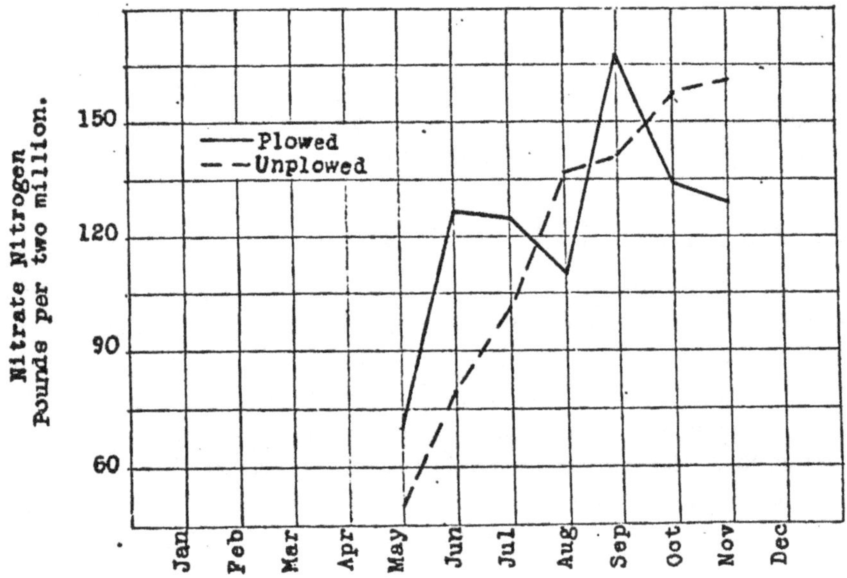

Fig. 5. Nitrates in fallow soil as influenced by plowing.

TABLE 2: Average corn yields (four years) on plowed as compared to unplowed soil.	
Treatment	**Bushels**
Plowed and prepared	16.33
Not plowed	10.26

The outstanding feature of these plats was the increase in nitrates after plowing. Each graph shows a distinct rise for the month in which it was plowed. Plat 2 shows a distinct rise in July; Plat 3 does not rise until August; and Plat 4 falls until August and then rises in September. These increases occur after all plats have come to about the same low level in June, or the month during the latter part of which the wheat was harvested. For the remainder of the season following the time of plowing, and for the early part of the next, the nitrate contents of these plats arrange themselves in decreasing amounts for the later dates of plowing. Plat 3 has an irregularity in the form of a high average for March, reaching 20.98 pounds on account of high nitrates in March, 1921. Severe dry weather during February and March of this year brought crystallized salts to the surface as a distinct bloom. This may have raised the nitrates, but apparently failed to do so in the other two plats. The March average of this plat for the four years following is but 15.7 pounds and would place the graph in more nearly expected order.

That the differences in nitrates were not inherent in the plats at the outset, but are the result of different dates of plowing and effects resulting therefrom, is shown by the nitrate content of these plats before any of them were plowed. Four samplings from May to the middle of July, when the first of the series of plats was plowed gave an average nitrate content as nitrogen for these plats as follows:

Plat 2..20.6 pounds
Plat 3..21.8 pounds
Plat 4..19.7 pounds

There was no greater variation within these three plats at any one sampling than 3.7 pounds of nitrogen per acre.

The low nitrogen in summer in the later-plowed plats is very probably due to the removal by weed growth. This was regularly heavy on Plat 4 before it was plowed in September. The crop of weeds already present in the wheat stubble at the time of wheat harvest was heavier for the later-plowed plats than for the earlier-plowed plats. The difference in weeds at harvest time between Plats 3 and 4 is shown in Fig. 1. This was the characteristic difference regularly for these plats, while Plat 2 plowed on July 15 seldom had any weeds, since none grew to form seeds to be replanted by the late plowing as was the case in Plat 4.

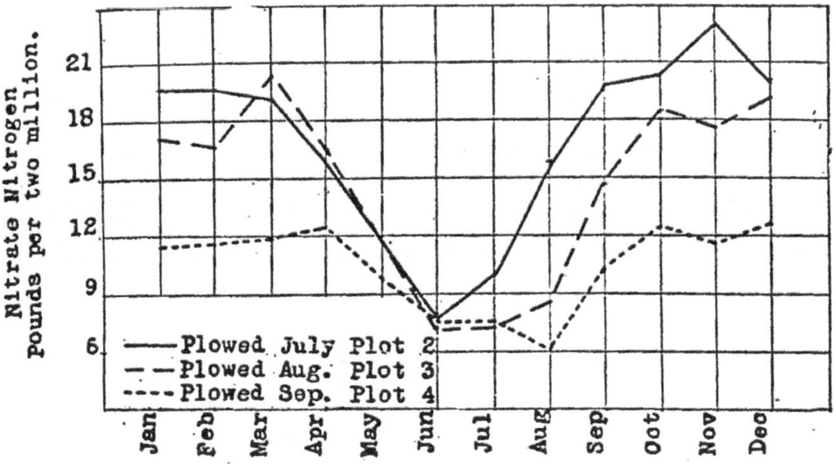

Fig. 6. Nitrates in soil in wheat as a result of different dates of fall plowing.

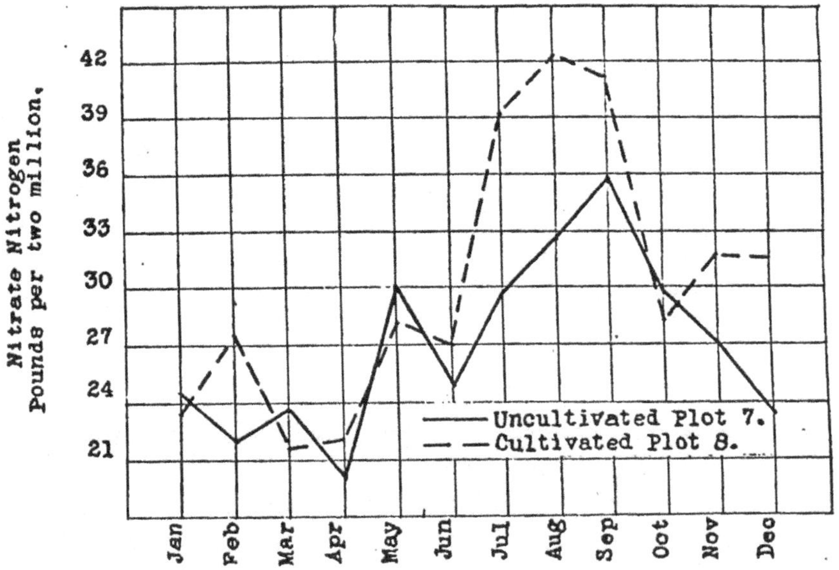

Fig. 7. Nitrates in fallow soil as influenced by ordinary surface cultivation.

The heavy weed growth on Plat 4 raises some question regarding the value of a catch or cover crop on this soil. Plat 4 had a catch crop of weeds growing for two months which was turned under in September to be nitrified later. The curves for nitrates fail to show, however, that this has contributed significantly, since this plat is never the equal in nitrates of the other plats with less, or no, catch crop, except when all three reach their lowest level in June. One might suppose, as is usually done, that the effect of the weeds has been to conserve the nitrogen so that an average figure might not show this effect. If this were true, the differences in nitrate content between the plats should become less as time goes on, to show the effect of this nitrogen conservation in comparison to two months of exposure of the bare soil in the early-plowed plat. Contrary to this supposition, however, the difference between the annual averages for nitrates in the early-plowed and the late-plowed plats for the last three years has been 5.5, 5.6, and 5.3 pounds of nitrate nitrogen per acre, respectively. This indicates that regular growth of weeds during the summer has not conserved nitrogen for its conversion into nitrate in later years, and the plat plowed in July to be left bare during the summer goes on producing more nitrates after five years than the plat with a catch crop supposedly saving nitrogen for future conversion.

The crop yields on these plats arrange themselves in the same order as for nitrate content or for time of plowing. Early plowing increased the yield, though this increase was not much greater for July than for August plowing. The data for the yields are given in Table 3 as bushels of grain per acre.

Though the nitrates have been higher regularly in the early-plowed plat, this showed no visible difference either in the fall growth of the wheat or the early spring growth. Later growth showed differences which became pronounced before harvest as shown in Fig. 2 for June, 1923. Since the rainfall curve for this locality suggests no shortage of soil moisture for wheat seeded in late September or early October, the influence of early plowing on wheat yield in this instance seems more probably to have been through increased nitrate production than through moisture conservation.

Influence of Surface Cultivation

That surface cultivation for three times as is ordinarily given corn during the growing season has an influence on nitrate accumulation is evident from Fig. 4 when one compares the graphs for Plats 9 and 10. Plat 9 was prepared in the spring; planted to corn; and cultivated with the regular eight-shovel cultivator three times during the season. Plat 10 received the same spring treatment but was scraped with a hoe to keep down the weeds in place of being cultivated. The nitrate content was regularly higher, though not significantly, in the cultivated plat than in the uncultivated. The average of five years for this difference in nitrates was only 0.63 pounds of nitrogen. The five-year average yields of grain for these two plats were 23.33 and 16.33 bushels, respectively, giving a difference of 7.0 bushels, or much greater than the difference in nitrates would suggest.

TABLE 3: Wheat yields in bushels of grain per acre as influenced by date of plowing.

Treatment	1921	1922	1923	1924	1925	5-year average
Plowed July 15	10.3	13.2	13.7	13.8	13.6	12.9
Plowed Aug. 15	10.3	14.1	11.3	12.4	15.5	12.5
Plowed Sept. 15	7.9	11.2	7.2	8.4	6.0	8.1

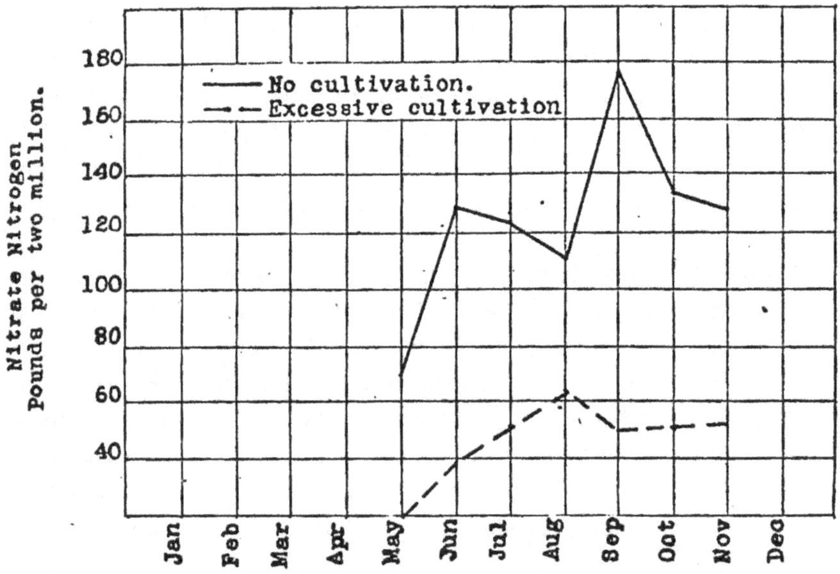

Fig. 8. Nitrates in fallow soil as influenced by excessive surface cultivation.

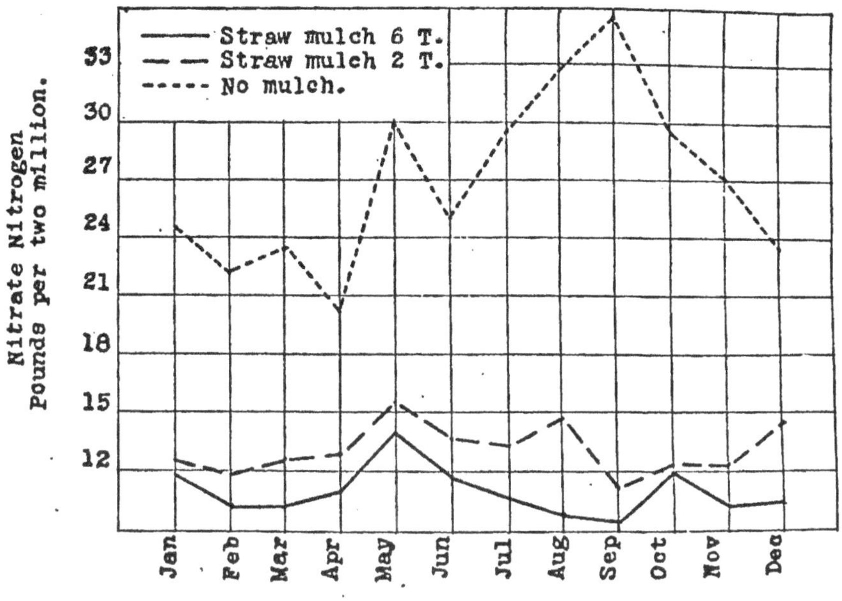

Fig. 9. Nitrates in soil as influenced by the straw mulch.

Plats 7 and 8 in this series were given the same treatment as Plats 9 and 10 but were uncropped. They also showed significant increases in nitrates as a result of cultivation with the difference becoming most marked after June, the time of the first cultivation. The data are given in Fig. 7. Two fallow plats in the series of 1917 to 1919 were similar in treatment, save that cultivation was given after every rain. Here the excessive cultivation gave a nitrate accumulation regularly below that in the scraped soil. The data for these two plats are given in Fig. 8.

The data indicate that the regular cultivation served to increase nitrate accumulation but that excessive cultivation served to lessen it. Doubtless the latter treatment kept so much of the surface layer continually dry to the exclusion of nitrate production that sampling to a depth of 7 inches failed to get enough of the regularly moist soil to give significant nitrates.

Effects of the Straw Mulch

That the straw mulch in applications heavy enough to exclude weed growth has a significant depressing effect on nitrate accumulation in soil had previously been reported.[2] In this study two uncropped plats were given applications of 6 tons and 2 tons of wheat straw, respectively. This was left on the soil until early the following spring when it was removed to allow the soil to dry preparatory to spring tillage. The lighter application was not enough to prohibit weed growth, consequently weeds were pulled to keep the plat clean. The nitrate nitrogen found in these plats is given as graphs in Fig. 9 in comparison with that of Plat 7, the uncropped, scraped plat without the straw mulch. These curves show clearly the depressing effect of the mulch in both applications, the heavier being the severer. No single treatment in the entire study showed such significant influence on the nitrate accumulation as the straw mulch. Only in May, after the old mulch had been removed and the soil plowed and remulched, was there any regular and significant rise in nitrates as an average. During single years, significant droughts increased them, but even these increased the total nitrate nitrogen content for the heavily mulched plats to more than 20 pounds per acre but once during the last five years. For over one-half of the determinations, the soluble nitrogen content of this plat was below 10 pounds per acre, or almost as low as that of the plat continually in grass.

The depressing effect of the straw mulch on nitrate accumulation has been shown[3] to be due largely to the increased moisture, which gives a bad physical condition with the exclusion of air to inhibit nitrate accumulation. This effect can be removed or produced in about two months by removing or applying the straw. This suggests the need for caution in selecting the condition under which the straw mulch is to be used.

Summary

Studies on nitrate accumulations in soils under the straw mulch and various kinds of tillage have shown that these treatments influence the nitrate content of the soil.

[2] *Albrecht, Wm. A. Nitrate accumulation under the straw mulch. In Soil Sci., 14: 299–305. 1922.*

[3] *Albrecht, Wm. A., and Uhland, R. E. Nitrate accumulations under straw mulch. In Soil Sci., 20: 253–267. 1925.*

The nitrates in an untilled soil, under sod, arranged by months are closely correlated with rainfall distribution of the year.

Spring-plowed plats are regularly higher in nitrates than those unplowed, though this increase may not take place immediately following the plowing. These differences are pronounced during the summer season and almost disappear during the winter.

Fall plowing has a distinct influence on nitrate accumulation as it is done in July or August. Significant nitrate increases follow after plowing for wheat, with greatest increases occurring in the soil plowed in the middle of July.

Weeds growing after wheat and plowed under in the middle of September seem to have no significant effect in raising the nitrate content to the equal of that in the soil without weeds and plowed early.

Ordinary shallow surface cultivation increases the amount of nitrates in the surface soil, but excessive cultivation lessens it.

The straw mulch has a decidedly depressing effect on the nitrate accumulation in the soil. This effect may be removed in about two months by removing the straw mulch.

CHAPTER 5

The Nitrate Nitrogen in the Soil as Influenced by the Crop and the Soil Treatments[1]

Introduction

The crop production on many soils depends largely on the supply of nitrogen these offer the plant. This supply is present mainly as organic matter nitrogen, which is in the plant residues, manure, and similar nitrogenous materials undergoing decay in the soil. During this decomposition the activities of the soil micro-organisms transform the nitrogen from the organic to the ammonia and nitrate or mineral forms, both of which are used by the plants. The amount present in the ammonia form is small, since in most cultivated soils this is quickly converted into the nitrate which is removed by plants, or leached out by the rains. The nitrate may also be consumed by microbiological activity, most noticeably so when excessive amounts of carbonaceous matter such as straw, for example, are turned under. Though the ammonia form may serve to supply nitrogen, the nitrate form is the main supply in cultivated soils, and plant growth generally increases as the supply of nitrate nitrogen increases. For this reason, the nitrate supply of the soil becomes of decided importance in crop production.

The study reported herewith was undertaken in order to become familiar with the nitrate supply of the soil during several years as related to the kind of crop, the soil tillage, the straw mulch, the fertilizer treatments and

[1] *W. A. Albrecht, "The Nitrate Nitrogen in the Soil as Influenced by the Crop and the Soil Treatments," Research Bulletin 250 (Missouri Agricultural Experiment Station, February 1937).*

other possible factors, as a partial help in solving the problem of providing sufficient "nitrogen turnover"[2] for higher crop production.

Plan and Methods of the Study

General Plan—This study deals with regular determinations of the nitrate nitrogen in the soil of a series of eleven plots of silt loam,[3] a rolling phase of Shelby silt loam, having different crops and cultural variations during six years, 1920–1925 inclusive, with different soil and fertilizer treatments along with different cropping practices during the succeeding seven years, 1926–1932 inclusive. The scheme of crops, of cultural, and of fertilizer treatments followed is given in Table 1.

During the last seven years there were added six nearby plots undergoing additional manurial treatments. Four of these six represented a corn-legume rotation, including (1) a legume crop, (2) a legume crop with residual fertilizer effects following corn, (3) corn given a legume green manure, and (4) corn given 300 pounds of 2-10-2 fertilizer in addition to green manure. Two of these six additional plots represented a corn, wheat and clover rotation on limed land. One was under the livestock system of farming, applying two tons of manure and 200 pounds of phosphate ahead of the wheat, and turning under clover ahead of the corn. The second represented the grain system of farming using 250 pounds of a 2-12-2 fertilizer on the corn and on the wheat with additional nitrogen as sodium nitrate on the wheat in the spring.

After the close of the nitrate studies, the eleven plots which had been under study for thirteen years, were given treatments of fine limestone, fine limestone and phosphorus, and fine limestone along with both phosphorus and potassium, and seeded to a red and sweet clover mixture. This permitted the observation of the clover crop as it reflected the lowered fertility levels suggested by the lower levels of nitrates at the close of the study.

---

2 Attention is directed to the term "nitrogen turnover." This term is used to represent the nitrogen in the course of transformation, by microbiological activity of the soil, from organic to other forms from which the plant obtains a portion. This term is used in distinction from the term "nitrogen supply" suggesting a stock of more or less permanent nitrogenous organic matter of the soil from which nitrogen is withdrawn for the plant needs.

3 This is a rolling phase of Shelby silt loam in a transition area approaching Bindley loam.

TABLE 1: Crops and Soil Treatments of Plots Under Nitrate Study					
Plot No.	1920–1925 Incl.		1926–1932 Incl.		
	Crop	Treatment	Crop	Treatments	
		Cultural		Cultural	Fertilizer
1	Corn	Unplowed Scraped	Fallow	Seedbed (a) Cultivated	
2	Corn	Seedbed (a) Scraped	Corn	Seedbed Cultivated	
3	Corn	Seedbed Cultivated	Corn	Seedbed Cultivated	N (b)
4	Fallow	Seedbed Cultivated	Corn	Seedbed Cultivated	N, Limestone (c)
5	Fallow	Seedbed Scraped	Corn	Seedbed Cultivated	Limestone
6	Fallow	Seedbed Straw Mulch (d)	Corn	Seedbed Straw Mulch	Green Manure (d)
7	Fallow	Seedbed Straw Mulch (e)	Fallow	Seedbed Straw Mulch (e)	Green Manure
8	Wheat	Plowed Sept. 15	Wheat	Plowed Aug. 15	N, P. K (f)
9	Wheat	Plowed Aug. 15	Wheat	Plowed Aug. 15	N, P
10	Wheat	Plowed July 15	Wheat	Plowed Aug. 15	N
11	Grass	Sod	Wheat	Plowed Aug. 15	

(a) This denotes plowing with a moldboard plow to a depth of 6 inches, disking and harrowing to produce a good seedbed in the spring, usually April.
(b) Ammonium Sulfate was broadcast at the rate of 50 pounds per acre ahead of corn planting.
(c) Limestone (mill run 10-mesh) was applied, broadcast at the rate of 2 tons per acre in the Springs of 1926 and 1927.
(d) The straw mulch was put on in April at the rate of 2 tons per acre 1920–1925, and 6 tons per acre 1926–1932. The green manure consisted of chopped sweet clover, applied at the rate of 2 tons per acre ahead of the plowing in Spring as the seed bed was prepared.
(e) Six tons of straw per acre.
(f) N represents ammonium sulfate 20 pounds; P represents super phosphate 150 pounds; and K represents muriate of potash 8 pounds, per acre, all broadcast in advance of wheat seeding.

Chemical Methods—Samples of at least nine borings per plot to a depth of seven inches were taken every two weeks (except when frozen soil prohibited) for nitrate determinations. They were thoroughly mixed and moisture determinations made on triplicate 100-gram samples.[4] These were then extracted with dilute acid, the extracts were boiled with alkali to remove the ammonia, and their nitrate content was then reduced with DeVarda's metal during distillation into standard acid for titration. All calculations were made as nitrate nitrogen per acre of two million pounds

[4] *The author is indebted to many students for help in taking the samples, and in making the chemical determinations. Acknowledgment of this help is hereby made to I. F. Wainscott, Helen Lemert Wainscott, Harold F. Rhoades, George Browning, Wm. T. Hearne, Albert Trowbridge, and T. Raymond Smith. The care of the plots by Mr. L. B. Stuckey is also gratefully acknowledged.*

of water-free soil. The extensive accumulation of data makes possible the presentation of only the average figures per plot by months or by years. These are reported in graphical rather than tabular forms and for the growing season in most cases rather than for the calendar year.

Experimental Results

Nitrate Accumulation Under Grass Followed by Wheat—The data for sod (6 years) show a correlation during the growing season, of the nitrate accumulation with the rainfall as shown in Figure 1, which gives the pounds nitrate nitrogen per acre and the rainfall in inches by months for the six years in question. The similarity of the curves is readily noticeable. There is an increase in nitrates with the increased rain and also with the rising temperature of the spring until the maximum is reached in May. A decline follows until August, probably because of the nitrate removal by the grass growth and decreasing moisture. A rise occurs again in September, when the increased moisture gives increased nitrates and the seasonal conditions may reduce the rate of its removal by slower grass growth. Regardless of the increases in nitrates in the spring and autumn, the figures never become very high. Grass evidently keeps the nitrate nitrogen exhausted to the low levels ranging between approximately six and twelve pounds per acre as the average of the six years on this soil.

When put into continuous wheat for seven years following the sod of six years, this plot showed no great nitrate content in consequence of turning under the sod and putting it into cultivation. Even though the figures for the first year following the sod were higher than those for the adjoining plot (Number 10) which had been in continuous wheat for the preceding six years, yet when averaged for the seven wheat years they are lower than was true for sod in all twelve months except in the two winter months of January and February, and the one summer month of August (Figure 1). Cropping this plot to wheat reduced the general nitrate level below that of sod, as shown in this figure where are given as a curve the monthly averages under the wheat for seven years following the six years of sod.

Nitrates Under Corn and Wheat—The results under sod and wheat raise the question as to whether the lower nitrate content under wheat is due to the particular ability of this crop to deplete it below the level possible by grass, or whether to the decline with time as it brings on nitrogen exhaustion. The influence of these two crops on the same plot, is shown in Figure 2, which gives the five year advancing averages of nitrates over

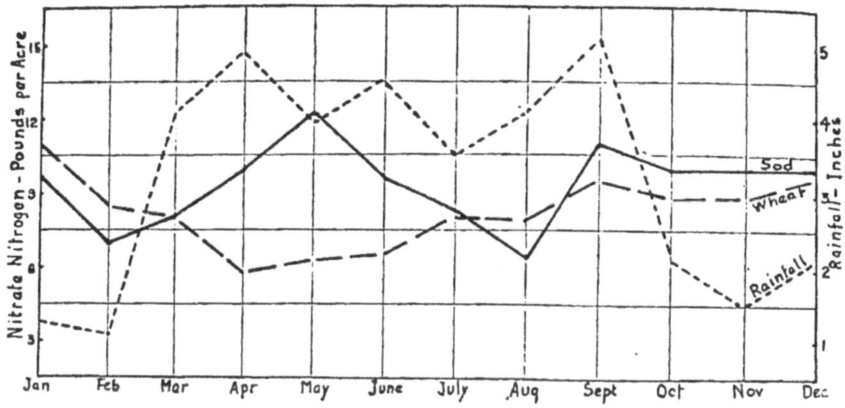

Fig. 1. Seasonal nitrate nitrogen content of soil in sod (6 years) as related to rainfall. Seasonal nitrate nitrogen in the same soil in wheat (7 years) following the sod.

the period including the six years of sod followed by the seven years of wheat. This curve fails to show any lower levels during the last periods to suggest a greater depletion by the wheat than was true for the grass in the first periods. It even suggests a possible rise during the last seven years. Since the nitrate level was so low during the entire thirteen years, and since the fluctuation is within a range of three pounds per acre, it is difficult to discriminate between the effects of the two crops as reducers of the nitrate supply.

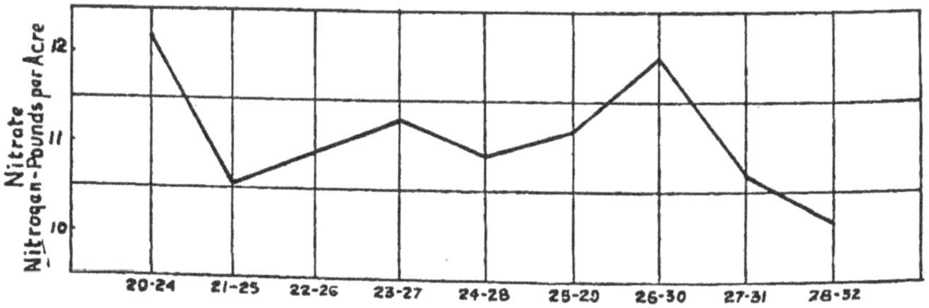

Fig. 2. Advancing five-year averages of annual nitrate nitrogen level in sod (6 years) followed by wheat (7 years).

The influences of the corn crop on the nitrate level in comparison to the influence by wheat is shown in Figure 3 in which the monthly averages for the growing season are given for these two crops. For wheat, the data are taken from three plots, one plowed in July, one in August, and one in September, all given no treatment for six years followed by seven years of fertilizer treatments consisting of additions of superphosphate, of ammonium sulphate and superphosphate, and of ammonium sulphate, superphosphate and potassium, respectively. During the seven years all these wheat plots were plowed on August 15. The data for the corn include two plots with regular cultivation and surface scraping but no fertilizer treatment for six years followed by regular cultivation and with ammonium sulphate broadcast at planting time in one plot during the succeeding seven years. The data for the three wheat plots and two corn plots are assembled as averages for the graphs in Figure 3.

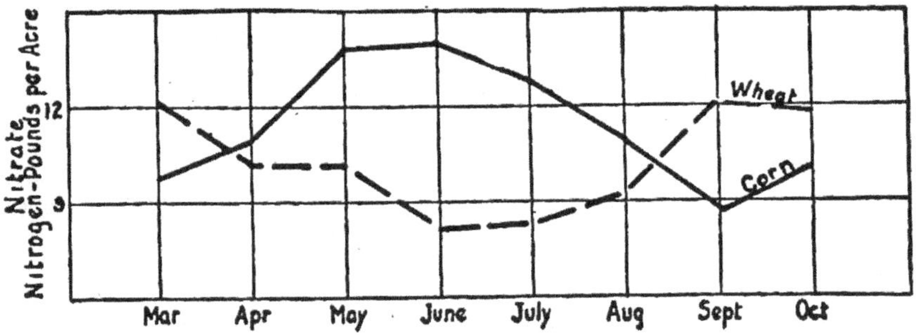

Fig. 3. Nitrate nitrogen levels in the soil during the growing season for wheat (3 plots) and corn (2 plots) during 13 years.

The seasonal curve for corn is the opposite of that for wheat. The former rises until May and June, declines to September and then rises again. The latter falls until June, rises until September and then falls slightly again.

Declining Levels of Nitrates Under Wheat and Corn—When the nitrates on these same plots are assembled as advancing five year averages of the season as given in Figure 4, the decided drop in nitrates with time is a most outstanding feature. There is also a marked divergence of these two curves with time because of a decidedly lower nitrate level under wheat than under corn. Since the cultural treatments of the soil under these two crops are so widely different, these effects by wheat can scarcely be ascribed wholly to the crop, and raise a question as to the effects of the tillage operations on the accumulations of nitrates.

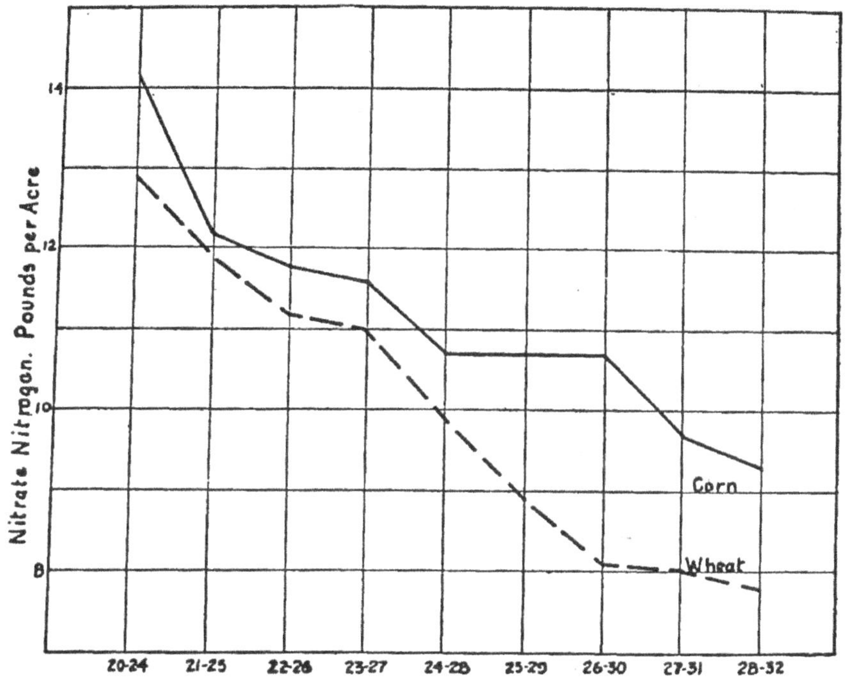

Fig. 4. Nitrate nitrogen levels in the soil in wheat and in corn as advancing five year averages of the season.

Nitrates in the Soil as Influenced by Tillage—*Turning with the Plow*—The tillage operations, both that of turning with the moldboard plow and that of surface cultivation, are the most influential soil treatments for nitrate increase. During the first six years, plots 1, 2 and 3 were designed to determine the influence of spring plowing and surface cultivation when all were grown to corn. As shown by the results given in Figure 5, the plowing of plot 2 increased the nitrate level markedly over plot 1, unplowed, especially during the growing season. From March to August the average difference was 2.16 pounds of nitrate nitrogen, while for the entire year this was 1.66 pounds. These differences reflected themselves in the yields of the corn crop. As a four year average this was 10.26 bushels per acre on the unplowed soil and 16.32 on that plowed and prepared but given no subsequent cultivation.

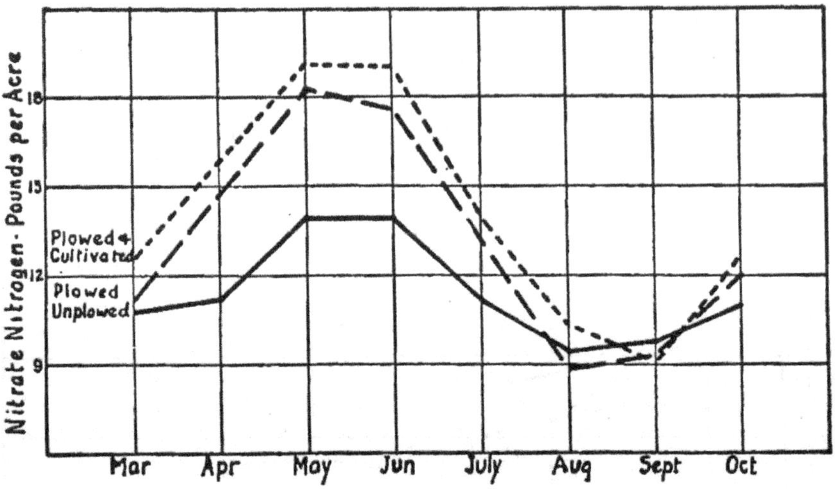

Fig. 5. Levels of nitrate nitrogen during the growing season in the soil under corn as influenced by plowing and surface cultivation. (6 years).

Surface Cultivation—Surface cultivation added to spring plowing was an additional means of increasing nitrate accumulation in the soil, as is shown by the curves (Figure 5). Scraping with a hoe removed the weeds in the unplowed and plowed plots, while the plot given the surface cultivation, commonly given to corn, produced an increase of less than 1.0 pound of nitrate nitrogen for the surface tillage. The yields of corn on the plowed and scraped plot and on that plowed and cultivated, namely 16.32 and 23.33 bushels, respectively, as the five year average gave a difference of 7.0 bushels corresponding to this small nitrate increase or influence by surface cultivation.

In the fallow plots, Numbers 4 and 5, which differed from each other only by this surface cultivation, the effect of this treatment was more marked. It manifested its effects particularly during the period June to October as given in Figure 6.

That these differences were due to tillage is indicated by the nitrate levels of the last seven years when the tillage treatments were the same for these different plots. For the latter period of the study, the nitrate nitrogen curves of plots 1, 2 and 3 are shown in Figure 7, and are all in different order than during the preceding six years as given in Figure 5. Plot 1, which was fallowed during the second period, was slightly higher than plot 3. Plot 2, continuously cropped and untreated, was decidedly the lowest of all three plots, and stood but little below plot 1, except during July

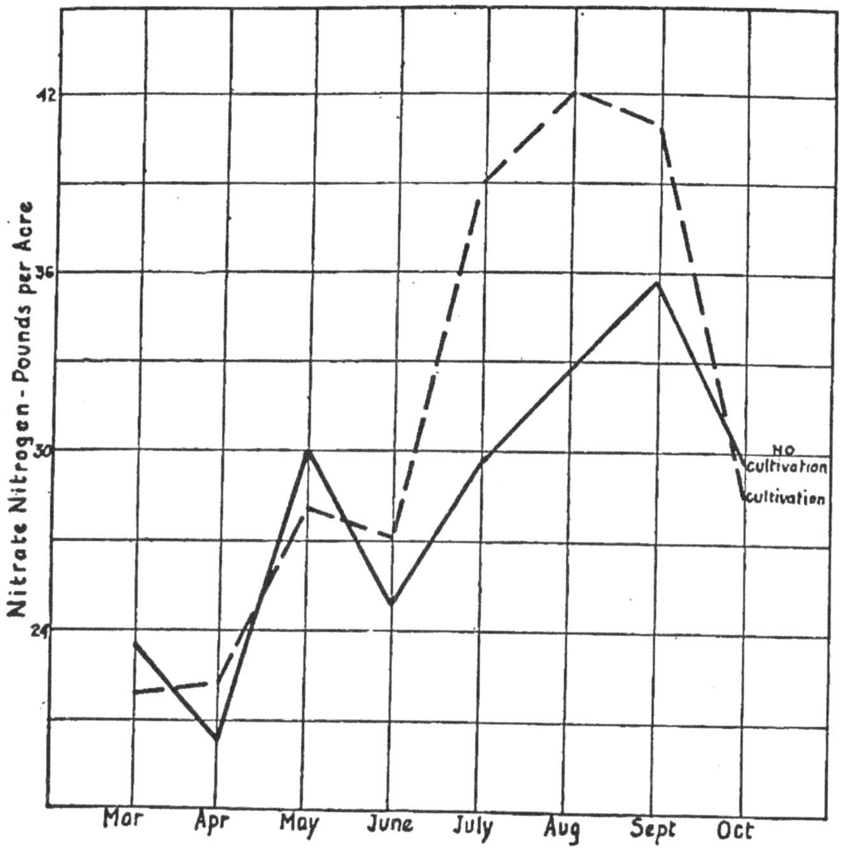

Fig. 6. Levels of nitrate nitrogen during the growing season in the soil under fallow as influenced by surface cultivation (6 years).

and August. During this latter period the plots reflected the effects of the fallowing and fertilizer treatments, when during the first six years they reflected the varied tillage treatments. It is interesting to note that the general level of the second period of seven years fluctuated between three and fifteen pounds of nitrogen per acre which was a level below that of the first six years. It points toward a decreasing nitrate nitrogen supply, as shown by graphs for plots 2 and 3 in corn included in Figure 4 as advancing five year averages. The corn crops on plots 2 and 3, which carried through both studies, and on plots 4 and 5 previously fallowed, are shown in Figures 8 and 8a, in one of the later years in the study.

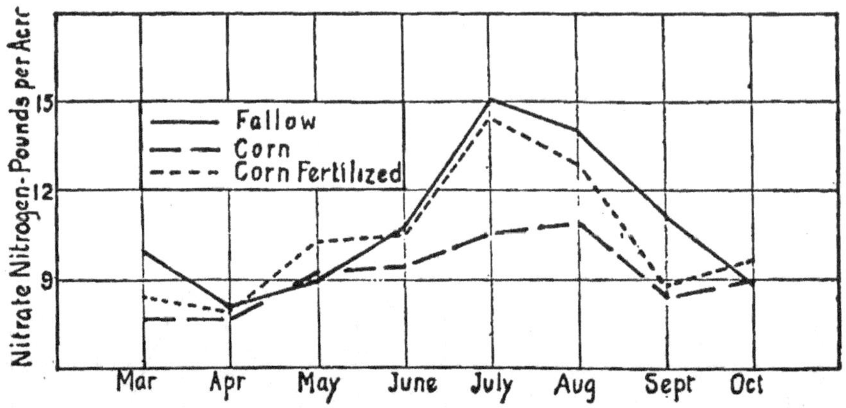

Fig. 7. Levels of nitrate nitrogen during the growing season in soil given similar tillage treatments (7 years) following varied tillage treatments (6 years).

Fig. 8. Corn crops in one of the later years of the study as they reflect the nitrate levels in the soil with different treatments. (No treatment, lime).

Fig. 8a. Corn crops in one of the later years of the study as they reflect the nitrate levels in the soil with different treatments. (Nitrogen, lime and nitrogen).

Date of Fall Plowing—The date at which the fall plowing is done for wheat emphasizes even more forcibly the influence of tillage on nitrate accumulation. These three plots were treated alike in all respects during the first six years, except that one was plowed on July 15, one on August 15, and one on September 15. The nitrate curves for the plots as given in Figure 9 reflect the importance of the date of plowing in giving added nitrate supply for the early October seeding of wheat. For this month the average figures for the nitrate nitrogen supply in the soil were 20, 18.5, and 13 pounds per acre according to July, August and September dates of plowing, respectively.

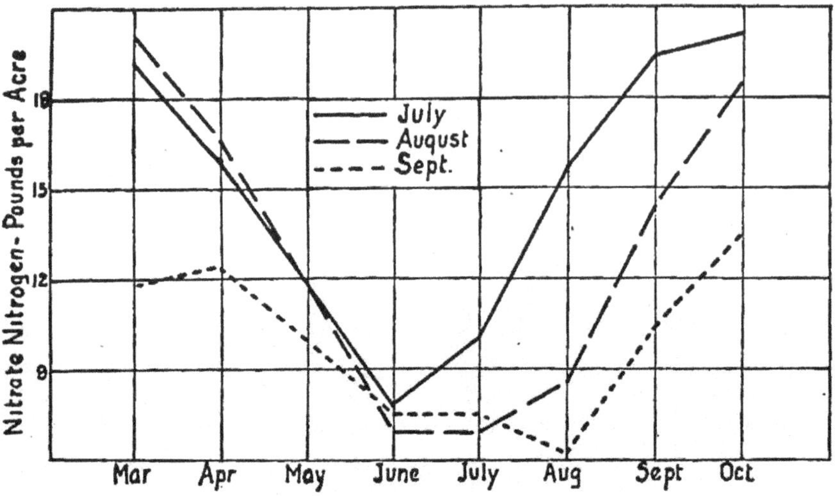

Fig. 9. Nitrate nitrogen in the soil in wheat as influenced by the date of fall plowing (6 years).

The increases in nitrates followed closely in each plot on the date of plowing as shown by the rise in the curve after the month in which the soil was turned. Counteracting this effect of plowing to increase the nitrates is the growth of the weeds on the unplowed soil to reduce them. More weeds grew on the plots as the plowing was delayed. This is shown in Figure 10. These differences in nitrates were reflected in the growth of the wheat crop illustrated in Figure 11, and in the five year average yields, which were 12.9, 12.5, and 8.1 bushels for July, August, and September plowing, respectively.

These same plots, Numbers 8, 9, and 10, when all were plowed at the same date (August 15) and given fertilizer treatments during the last seven years fail to reflect significant differences in nitrate content. The plot previously plowed in July was fertilized with nitrogen in the ammonium sulphate form, that plowed in August was given nitrogen and superphosphate, while the third one previously plowed in September was given nitrogen, phosphorus and potassium. Alongside these three plots, wheat was also grown with no treatment on the plot, Number 11, previously kept in sod. These treatments do not reflect themselves as differences in the nitrate curves given in Figure 12. These curves are all very similar and at noticeably low levels, fluctuating between six and ten pounds of nitrogen per acre. The monthly averages for seven years, when added for the growing season March to October inclusive, were 62.6, 63.2, 61.5 and 60.6 pounds of nitrate nitrogen for plots 8, 9, 10, 11, respectively.

Fig. 10. Differences in weeds in wheat stubble in July as a result of difference in time of fall plowing for the crop. (August 15, September 15, reading downward).

Fig. 11. Differences in the wheat crop in response to different dates of plowing for the crop. (July 15, August 15, September 15, reading downward).

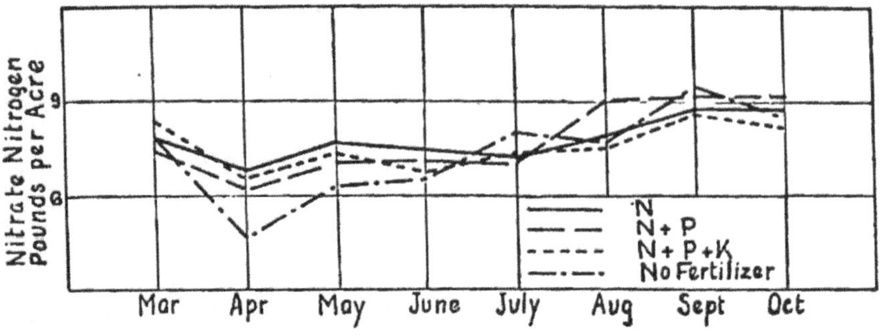

Fig. 12. Levels of nitrate nitrogen during the growing season under wheat given similar tillage but different fertilizer treatments.

Fertilizer Treatments and Nitrates—In case of both wheat and corn plots which reflected the influence of dates of plowing and types of tillage, respectively, in the first six years, the wheat plots showed little influence by fertilizer treatments during the second period and the corn plots, which previously reflected a distinct effect of tillage on nitrate accumulation reflected but slight effect of fertilizers on the nitrate supply. The influence of the fertilizers on nitrate residue seems insignificant, but it is impossible to say whether this low residual supply may not be due to a greater ability of the fertilized plots to exhaust the soil of this form of nitrogen. These corn plots maintained higher nitrate level at eight to fifteen pounds per acre in the last seven years than was the case of the wheat plots at six to ten pounds of nitrate nitrogen per acre. It is possible that the difference is due partly to the different abilities of these two crops to exhaust the nitrate nitrogen to a low level and to the additional cultivation given in the case of the corn. In spite of the nitrogenous fertilizers added, there was a decline of nitrates in the soil in the case of both crops as shown in Figure 4.

Straw Mulch and Nitrates—The detrimental influence of the straw mulch on the nitrate level is as pronounced as perhaps any single effect by a treatment. Two plots, Numbers 6 and 7, were uncropped during the first six years with a wheat straw mulch of two tons and six tons, respectively, applied in April after seedbed preparation. The low nitrate level, going with the heavier straw mulch, in particular, was very noticeable during the first period as shown in Figure 13. In order to test the possibility of raising this by adding a nitrogen carrier, there was turned under each spring during the last seven years, chopped sweet clover, equivalent to the rate of two tons per acre. One of the plots, Number 6, was also planted to corn.

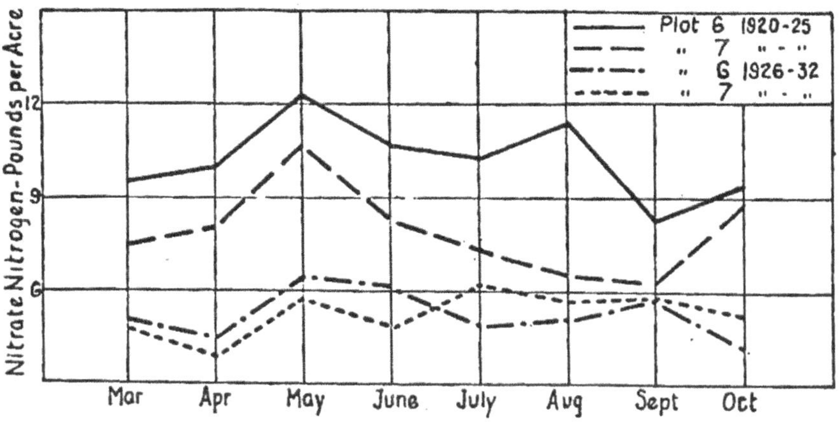

Fig. 13. Nitrate nitrogen levels during the growing season under a straw mulch on fallow (6 years), and with green manure treatment on both fallow and corn (7 years). (1920–25. Plot 6, fallow, 2 tons straw mulch. Plot 7, fallow, 6 tons straw mulch. 1926–32. Plot 6, corn, green manure and 6 tons straw mulch. Plot 7, fallow, green manure and 6 tons straw mulch).

The outstanding observation is the fact that during the second period of seven years the nitrate levels were still lower than during the first, regardless of the addition of the sweet clover. The plot under the mulch and growing the corn was slightly lower in nitrates from July to October in the average of the last seven years than the fallowed, mulched plot. But this difference seems insignificant since the two curves in the later period are very similar. In the curves of the advancing five year averages, Figure 14, these two plots are almost alike after the first three five-year periods. Their difference at the beginning was due to the lighter application of two tons of straw on plot 6 in contrast to six tons on plot 7. Further, both plots dropped in their nitrate levels whether under fallow or under corn, so that their lowest points were slightly below the low point of continuous unmulched corn, and slightly above the lowest point of continuous wheat (Figure 4). The ability of plot 6 to grow corn is illustrated in Figure 15, which may be contrasted with the crop in Figures 8 and 8a, showing the crop on unmulched soils at the same time.

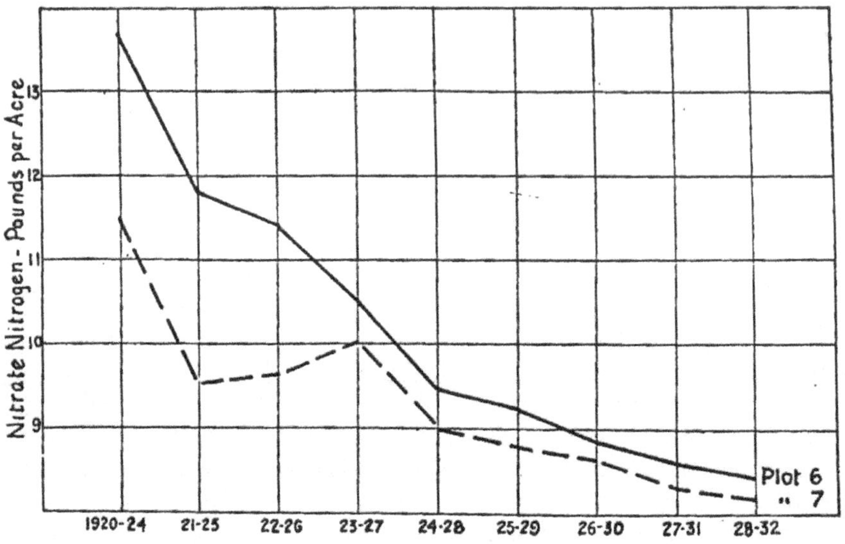

Fig. 14. Levels of nitrate nitrogen during the growing season as advancing five year averages for fallow soil, unfertilized or given green manure and under straw mulch. (Plot 6, fallow under 2 tons straw mulch followed by corn with green manure and 6 tons straw mulch. Plot 7, fallow under 6 tons straw mulch followed by green manure and the same mulch.)

Fig. 15. The corn crop in one of the later years of the study on soil given green manure and straw mulch.

Determinations of ammonia nitrogen in these mulched soils at the same time that the nitrates were measured did not suggest excessive accumulations of this form of nitrogen. Rather, the low level of nitrates was due to the addition of organic matter of a wide carbon-nitrogen ratio (straw) and the consequent consumption of nitrate nitrogen by the soil micro-organisms. No complete report of the ammonia studies is included.

Crop Rotations and Nitrate Accumulation—Where the corn and legume rotation was the system of management, the nitrate accumulations during the last seven years were on the same general seasonal levels as where corn was cropped singly with the different fertilizer treatments. This was true whether the corn crop in the rotation was given 300 pounds of a 2-12-2 fertilizer, or whether it was omitted. The similarity of the seasonal nitrate levels in the rotation with these treatments as given in Figure 16 to those of plots in continuous corn is evident by comparing this figure with Figure 5. There is a suggestion that the activity was somewhat different in July where no fertilizer was used, since the residual nitrogen as nitrate was higher in this month. There was a significant difference in the two curves only at this time, though the fertilized plot was higher in the early season and lower in the later season. It gives a suggestion that the fertilizer helped nitrate production early in the season, and encouraged more complete consumption of the nitrates by the crop later.

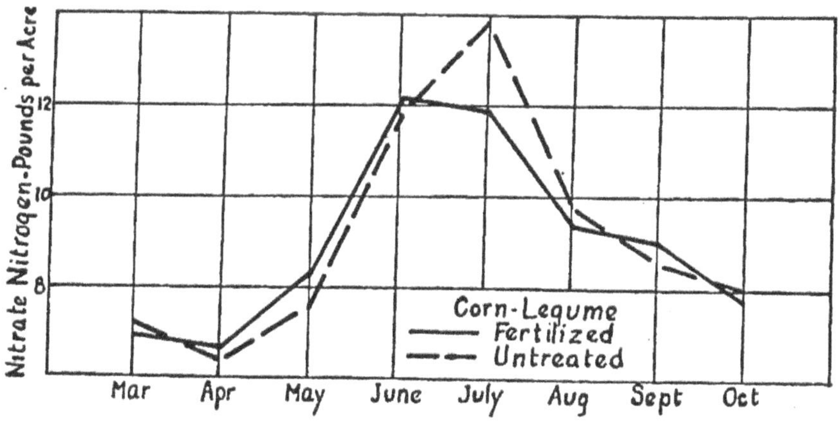

Fig. 16. Seasonal nitrate levels in soil under a corn-legume rotation with and without fertilizer treatments.

Since the crop yields in the rotation were larger than those in continuous cropping to corn, it is evident that this soil was delivering more nitrogen to the corn in this rotation. However, in both cases the corn was exhausting the nitrates to the lowest levels and since these levels of remaining nitrates are those measured, these residues of nitrogen then may well correspond, irrespective of the larger amounts that may have been produced and removed from the soil by the higher corn yields in the case of the rotation.

In the rotation of corn, wheat and clover with the livestock and grain systems of farming, there again the two plots were very similar in their behavior. Since corn occurred only three times in each, and the wheat and clover each only twice during the seven years, it is scarcely advisable to separate these data according to the crops. Consequently, the data are plotted as averages for the seven years during the growing season as shown in Figure 17. The close similarity of these to those in the corn-legume rotation and to all the other corn crop curves is evident. Again, larger crops were produced but since only the residual nitrate at any one time was measured, no account is taken of the total nitrogen transformed as nitrates.

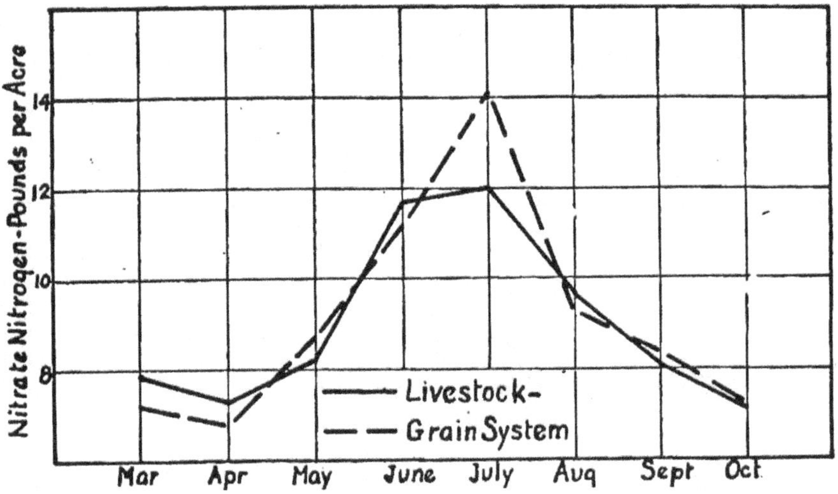

Fig. 17. Nitrate levels in the soil under corn, wheat, clover rotation for the livestock and grain systems of soil treatment.

As was true for the corn-legume rotation, likewise in the three year rotation there was a higher accumulation of nitrates in July in one case than in the other. Here this occurred in the grain system of farming with its fertilizer treatment, while in the former case it was in the unfertilized soil. No significance can be attached to the fertilizer treatment as the cause of this condition, since fertilizer was applied in the grain system where the nitrates were high in July and was omitted in the other case where they were high in this month. Whether this larger amount of nitrate during this month in these two cases has any significant causal factor cannot be discovered from any evidence.

Soil Fertility Exhaustion Demonstrated—The advancing five year averages of nitrate levels show such a decided decline, (Figures 4 and 14), that one can interpret such only as a lowered ability of this soil to deliver nitrates. This decline is more marked when the seasonal cycle of nitrates for succeeding years or some few of them in the succession of years are plotted together as for corn in Figure 18. Not only are the levels of nitrate lower with succeeding years of continued cropping, but the high point of the year is reached at a later date in the season. During the first five year period this high point under corn was reached in May, during the next five year period selected for two years later, this high point came in June, then in the next period later by three years than the second it advanced to July and in the last five year period to August. Thus, as the cropping continued under continuous corn, this soil was able to deliver its maximum amount of soluble nitrogen only at a successively later date in the season and at successively lower levels. Curves plotted similarly for the soil in wheat and for that in fallow demonstrate lower levels corresponding to those for corn and emphasize the declining ability of this soil to deliver nitrates.

That this decline is in the fertility as well as in nitrates is further indicated in the results from the seeding of clovers on these soils after the study of nitrates ceased. These plots were treated with limestone, and this combined with phosphorus and with both phosphorus and potassium and seeded to a clover mixture. On the plots grown continuously to wheat previously, scattered stands of improved numbers of plants agreeing with the increased additions of fertilizer were obtained. But these numbers were still too insignificant to hold the clover crop. Only red clover plants were in the stands. No sweet clover plants were obtained in this area formerly in wheat.

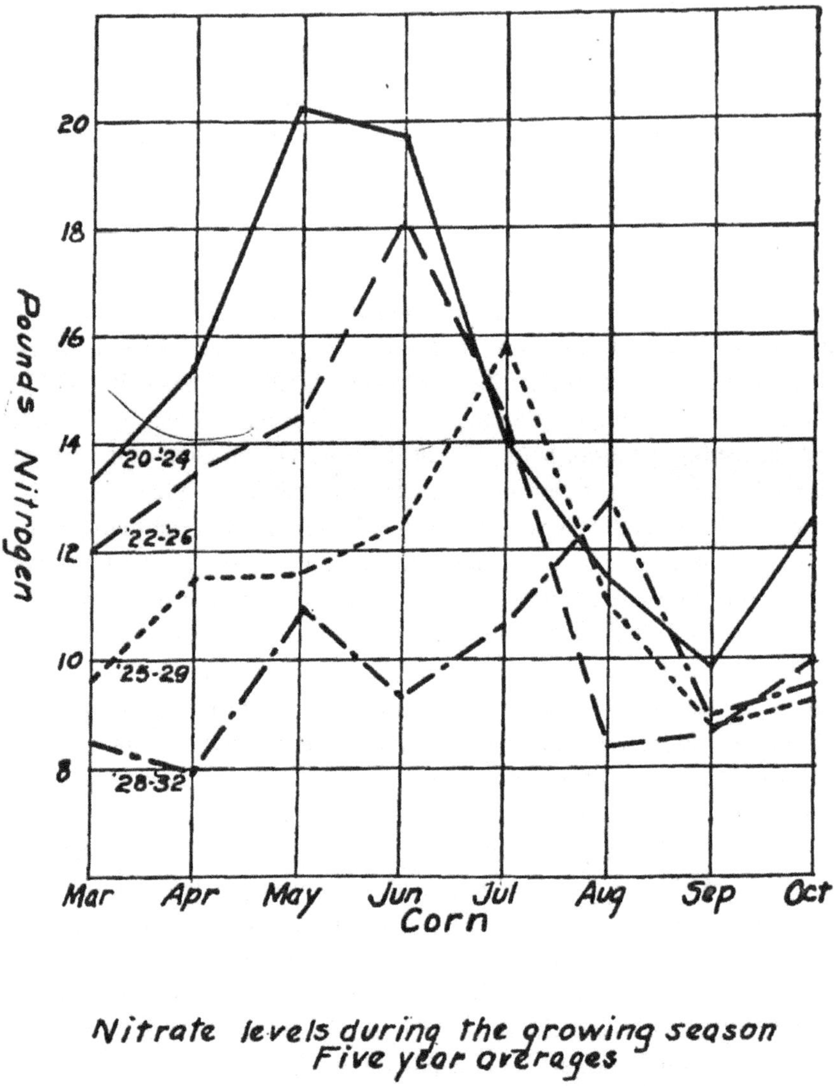

Fig. 18. Seasonal levels of nitrate under corn as averaged for different five-year periods showing the declining nitrate levels and later seasonal maxima with continued cropping.

Fig. 19. View of the plots on which the varying levels of nitrate nitrogen were revealed by differences in clover stands and other crops. In the foreground is seen the sweet clover stand after corn; in the background, sweet clover failure after wheat.

On the plots formerly fallowed, mulched, or grown to corn, the situation was quite different. There was no red clover, but a good stand of sweet clover was obtained. This improved with the limestone and fertilizer additions. The influence of lime additions was pronounced, especially those additions of limestone made during the period of nitrate study. The plots formerly fallowed likewise reflected their higher fertility in better sweet clover because they had remained uncropped. The sweet clover crop on these plots was of such density and growth as to warrant its consideration as a fair crop in contrast to that on the former wheat plots which could in no way be so considered. Some suggestions as to these differences in fertility and clover crops may be obtained from the illustration Figure 19 representing the spring condition of the crop in the second year after the clover seeding, and two years after the close of the nitrate study. Figure 20 gives further detailed differences at the same time for one of the plots (Number 7) which was fallowed and given a straw mulch during the thirteen years;

and one (Number 10) which had been in wheat continuously with August plowing and no fertilizer treatment for six years followed by similar plowing date and the additions of nitrogen and phosphorus as fertilizers. These clover crop differences reflect clearly the differences in fertility which were also reflected in the different levels of nitrate accumulation.

Fig. 20. Difference in crops resulting from clover seeding. Sweet clover stand on plot (No. 7) fallowed and mulched in contrast to bracted plantain (Plantago aristata) on plot (No. 10) formerly in continuous wheat.

Summary

In this study, measure was taken regularly of the accumulation of nitrates in the soil under different crops, different methods of tillage or cultivation, different fertilizer treatments and under the straw mulch. Such determinations do not measure the amount of nitrates produced or removed, but rather the nitrate level at which the production of nitrates is temporarily balanced against their consumption by the plants as the crop above the soil or by the nitrate consuming microorganisms within it. They measure, however, the lowest level to which the supply in the soil can be, or is, exhausted by the crops or by leaching, and the heights to which the accumulation may mount against, or in absence of, these reducing forces.

That the crop is a significant factor in removing the nitrates is shown by the lower nitrate accumulation in the cropped soil in contrast to that in the fallowed soil. Without the crop the seasonal nitrate nitrogen levels fluctuated from approximately 21 pounds at the lower level to 42 pounds as the upper figure. For the crops, their exhaustion of the nitrates was very similar whether grass, corn, or wheat. The similarity was greatest between wheat and grass whose average lower levels were the same at 6 pounds and upper levels also alike at 12 pounds.

For corn the two corresponding figures were each two pounds higher at 8 and 14 pounds, respectively. These facts demonstrate clearly that nitrates accumulate in the fallow soil and are removed by the crops to a degree closely similar for the same soil regardless of the crop.

The accumulation of nitrates follows seasonal conditions closely. With the advent of spring the nitrates in the soil increase and show that this process like any other biochemical soil process is subject to the moisture level and the prevailing seasonal temperature. It is set to agree closely in its activity with that of the growing crop.

Soil tillage is a beneficial soil treatment because of its effect in bringing about an increased supply of available nitrogen or nitrate for the crop growing on the soils. Turning the soil with the moldboard plow is the foremost tillage performance to increase nitrate production. Surface cultivation likewise stimulates nitrate production. Plowing is effective whether the soil is tilled in the spring, April, or in the later summer months of July, August and September. These differences in nitrates would lead one to believe they suggest that plowing and tillage are not fully appreciated for their effects on this process of transforming soil nitrogen into the nitrate or soluble nitrogen form.

Fertilizers did not bring about any great change in the level to which nitrates were exhausted. Their effects were generally unnoticed in terms of the measurement of the nitrate nitrogen supply on the soil though there were suggestions that they encouraged early season nitrate accumulations and aided in the late season nitrate removals. These were but suggestions by small differences in no way commensurate with the differences in the crop yields from fertilizer use. This study did not measure the possible increase in total nitrate production from fertilizer use, but only the accumulated level, hence such items as the influence of fertilizer on total nitrate production could not be determined. Perhaps heavier additions of fertilizer are necessary to raise the level above that found in these trials.

The most detrimental effect to nitrate accumulation in a fallow soil in these studies was manifested by the straw mulch. Under the application of six tons of straw per acre, the fallow soil failed to accumulate nitrates above the levels common in soils growing crops. This effect was not overcome when nitrogenous green manures were turned under in the form of sweet clover. The growth of corn in the straw mulched soil was, however, the equivalent of that on unmulched soils where higher accumulations of nitrate occurred.

Whether continuous cropping or crop rotation was practiced, the level of nitrate nitrogen under the same crop was but little disturbed as a general average. The nitrates under corn continuously were closely similar to those under corn alternated with a legume crop; to those under corn in a three year rotation handled as a livestock system of farming, and to those under the grain system of soil management. This suggests that these different systems like fertilizing are not reflected in raising or lowering the levels at which there is the equilibrium between production and exhaustion of the nitrate in a soil at any one time. In all these cases of different cropping systems the soil was holding the nitrate level to about the same number of pounds per acre. This suggests that the crops were reducing it to the limit which was a constant for this soil and the particular crops, whether in a rotation or continuous sequence.

Perhaps the most noticeable feature of the nitrate levels observed in this soil was the decline of the levels with time. Under the wheat or under the corn, and likewise under the fallow soils, both mulched and unmulched, the nitrate levels observed were lower as the years of the study advanced. This points forcibly to the decline in ability of this soil to deliver nitrates early in the season and at high levels as cropping continued. These levels had fallen so low that it was difficult to establish sod legumes such as red clover and sweet clover on these soils even with an application of finely ground limestone and commercial fertilizer. These declines represented reductions in levels as much as 50 per cent and should direct our attention to the significant speed at which the soil's power to produce nitrates is being depleted when such reductions occur within a period as brief as 13 years.

CHAPTER 6

Methods of Incorporating Organic Matter with the Soil in Relation to Nitrogen Accumulations[1]

Summary

The careful analyses of soils sampled annually where two and one-half tons of clover, corresponding to 106.2 pounds of nitrogen per acre, were turned under each year and where this corresponding amount was applied on the surface and turned under the following year, show that nitrogen accumulated only slightly more rapidly in the latter than in the former. This difference was relatively small, amounting to about 3 per cent of the nitrogen applied. The close agreement between the results of the two treatments, especially the greater amount of residual nitrogen in the soil where clover was put on the surface, suggests that this surface application doesn't volatilize any nitrogen, nor does it have any loss; at least not more than where the clover is incorporated into the soil.

The study suggests that as a general average where clover is used for soil improvement in this manner, about one-third of the annual application may be considered as contributed to the residual or insoluble nitrogen supply in the soil while two-thirds are mobilized or transformed into soluble forms. With these heavy annual applications of clover, the soil supply of nitrogen was raised above its initial content at a rate amounting to more than 20 per cent of the annually applied nitrogen. Its effects in offsetting declines of nitrogen in the soil, added to this, make its effects in building

[1] *W. A. Albrecht, "Methods of Incorporating Organic Matter with the Soil in Relation to Nitrogen Accumulations," Research Bulletin 249 (Missouri Agricultural Experiment Station, December 1936).*

up the soil nitrogen approach the rate of one-third of the nitrogen applied annually.

These results were obtained on a shallow surface soil underlain by a plastic clay subsoil. Slight increases in the upper subsoil content of nitrogen resulted where significant increases occurred in the surface soil. The low initial content of nitrogen in the surface and subsurface soils may have been a factor in the effects obtained during the 15 years. Since many soils of the state are of similar profile and of no greater nitrogen content, these results suggest that in much of the state heavy clover additions can build up the nitrogen reserve in the soil. Though such repeated heavy applications are not common practice, these results indicate the effects which might be expected from the application rates here used, when the intervals between applications may be three or four years as is more common in practices aiming to maintain the soil.

Introduction

The incorporation of green manure by plowing it into the soil is the commonly recommended method in soil management. How effective legumes may be in soil improvement, when they are left to accumulate on the surface, has not been fully determined. Different biological conditions obtain in these two cases. The absence of sunlight within the soil as compared to its sterilizing effect on the surface is a distinct biological contrast. Moisture which is relatively abundant in the former case would favor decomposition, while the alternate wetting and drying in the latter case, with only short periods of significant moisture, would seem to militate against decomposition. Any volatile materials produced during intermittent and interrupted surface activities would quickly escape, but would readily be absorbed by the soil when produced within it.

By turning under green manures during the usual seasons of plowing, such as spring or fall, little consideration is commonly given to the fitness of the chemical composition of the green manure as a nitrogenous fertilizer, particularly its carbon-nitrogen ratio as determined by its growth stage. Possibly legumes intended for soil improvement can be left on the soil surface, or slightly disked in, with significant contribution to the nitrogen supply in the soil. If this can be done independently of the costly operation of turning the surface soil with the plow, green manure additions to the soil may be made more cheaply. The following study was undertaken in order to learn something about the differences in effects on the nitrogen content of the soil when red clover was turned into the soil promptly as compared to its application on the surface for much later incorporation.[2]

Plan of Experiment

Soil and Plot Treatment—The soil used in this study was a phase of Shelby silt loam. It was in a wheat crop when the experiment was begun and had been in general crops preceding. It was decidedly silty with a low content of clay or colloidal material in its surface. At the beginning of the study the untreated plot had a pH of 5.75 and at the close the figure was 5.50. The plot which was limed in 1929 with no further treatment had a pH of 6.30 at the end of the samplings.

Two series (North and South) of three plots each with dimensions of 6 feet by 6 feet 8 inches or .0009182 acre, were laid out with board curbings around them to prevent erosion and to give specific limitations from which measurements could be taken for accurate location of samplings. They were covered by framed screens of a mesh no larger than the mesh of fly screen, fitted accurately on the curbing to prohibit wind effects as removals or additions. One series of plots was laid out in 1917 and the other in 1918. Both were kept fallow and any weeds that started were pulled and left on the plots.[3]

In late June each year, or about the season when the regular cutting of red clover is made for hay, the plots were spaded to a guarded depth of seven inches. Dried and chopped red clover was spaded into the west plot of the three in each series or incorporated at the rate of 2,082.4 grams, which was the equivalent of two and a half tons per acre. This same amount of red clover was scattered on the surface of the east plot after it had been spaded. Any remnants of organic matter left on the surface from the application of the previous year were turned into the soil. These remnants were of small amounts as early, futile attempts to recover and measure them indicated. The central or middle plot was given no organic matter addition with the spading. No further tillage or soil disturbance occurred beyond

---

[2] *This problem was suggested by Professor M. F. Miller, in connection with the larger project of biological nitrogen fixation in soils.*

[3] *No legumes occurred on the plots as weeds, hence no nitrogen additions were made by nitrogen fixation through this means. The weed flora varied with years. It included such plants as shepherd's purse (Capsella Bursa pastoris), yellow sorrel (Oxalis stricta), purslane (Portulaca oleraceae), and mouse eared chickweed (Cerastium vulgatum). It was a common observation that more weeds grew on the clover treated than on the untreated plots, and that its application on the surface was more conducive to weeds than its incorporation into the soil.*

that occasioned when a few weeds were pulled. In 1929 the plots of the South series were given limestone of 30 mesh fineness equivalent to three tons per acre.

Sampling and Soil Sample Preparation—Before the plots were spaded each year, nine samplings per plot were made, including the surface seven inches by means of a two-inch tube passing through a metal plate guard, and the subsurface five inches by means of an auger passing through the tube. The location of the samples each year was determined according to a scheme in which there was a shift of four inches in one or the other direction paralleling the sides of the plots. This means was used to obviate sampling into subsurface holes of previous samples which had filled with surface soil. The nine samples of the surface soil were combined (likewise those of the subsurface), dried, ground in a ball mill and then a portion of about 200 grams of each put through a Braun pulverizer to reduce it so that it would pass through a 100-mesh sieve. This portion was used for chemical analysis, while part of the larger remaining sample was put into storage for reference.

A portion of the clover hay, which was usually obtained from other nearby experimental plots and dried in the screened shed before it was chopped into one inch lengths, was also taken as a sample for chemical analysis. Due to differences in moisture and maturity in the clovers, a rather wide variation in the annual application of nitrogen occurred. This ranged from 80 to 124.8 pounds per acre, with an average of 106.2 for the many applications.

Chemical Analyses of Soils—After the samples had been collected for three years, analyses were made for their total nitrogen. Since analyses of this type involve errors that may be large in relation to the variable that is to be measured, all later samples were held for analysis at the same time, using the same reagents, the same standard acids, the same chemist,[4] and uniformity in all conditions as far as possible. Ten gram samples of the air dry, 100 mesh soil were weighed, oven dried for ten hours at 107° C., weighed again, transferred to 800 cc. Kjeldahl flasks and digested to a gray white condition by 25 cc. of concentrated sulfuric acid given copper sulphate and sodium sulphate. They were then diluted, made alkaline and distilled into standard acid. This was .0714N, equivalent to 1 mgm.

[4] *The author is indebted to the painstaking accuracy of Mr. Fred Stonner, who through N.Y.A. aid made the chemical determinations under the author's direction and supervision. Likewise to Albert Trowbridge, George Browning and other student workers who aided in collecting and preparing the soil samples.*

of nitrogen per cc, and was titrated with alkali of similar normality from supplics sufficiently large to serve for the entire lot of samples. Blank distillations of the reagents were made regularly and corrections applied accordingly. Duplicate titrations varied no more than .1 cc. in the majority of cases, in an attempt to hold the variations to .2 cc. Duplicate determinations would thus differ by 20 pounds per two million with the former figure and 40 pounds for the larger variation. Repeat determinations were made in a few cases when variations in duplicates were considered beyond toleration. All calculations were made on the basis of two million pounds of water-free soil.

Results

The data as annual analyses are assembled in complete form by plots, by soil strata, by treatments, and by years in Table 1. These are presented graphically as a scatter diagram in Figure 1. As a means of lessening the fluctuations, they have been calculated as advancing three year averages with the first data under the year 1920, which was the third year including 1918. These averages are assembled in Table 2. The data of 1917 are omitted for the North series in order that the dates may be more comparable. These advancing average data are given graphically in Figures 2 and 3 for the surface soils, and in Figures 4 and 5 for the subsurface soils of the North and South series, respectively. As a measure of the changes in nitrogen content by the treated plots, the arithmetic mean of the three year advancing average figures of the two untreated plots (surface soils only) for the various years was subtracted from the corresponding figure for those treated. These data are assembled in Table 3. The average nitrogen content of the surface soils of the two series of plots are given graphically in Figure 6. The increases in surface soil nitrogen where the clover was incorporated and where it was applied on the surface, as compared with the untreated soil, are given in graphs in Figure 7. Since there seemed to be no significant effect from the lime treatment on the surface soil of the South series as shown by the curves after the year 1929, this comparison was disregarded in handling the data.

As a convenient general survey of the nitrogen changes in the plots, the data have been assembled using the first three-year average figure for that at the outset, and the last three-year average figure for that at the close of the study. The difference between the two may be taken as the gain or loss in residual nitrogen. Included also are the averages of all the analyses made of the samples taken annually during the period. These data are assembled in Table 4.

Nitrogen Supply in the Untreated Surface Soils—As for the untreated surface soils, the fallowing treatment during the period gradually depleted their supply of total nitrogen. There were some periods of variation but during the time as a whole it is evident that the nitrogen supply in the soil was becoming less as shown in Figures 1, 2 and 3. If one takes the final three-year average figure as contrasted with a similar period for the outset, as given in Table 4, the loss during the 15 years under study was 115 pounds per acre.

The situation as to nitrogen loss would have been little different had the soil been cropped and all the crops removed, since the nitrogen taken by crops is that in the soluble form, and would be leached were it not taken up by the plant roots. Crop removal leaves roots that represent a part of the soluble or leachable nitrogen returned in the soil in the organic and less soluble form. Other than for this item, the conditions under fallow and cropping would be similar with respect to nitrogen removal.

Nitrogen in the Surface Soils Receiving Clover—The most pronounced feature of the tables and figures is the increase in total nitrogen of the surface soil whether clover was incorporated or left on the surface and turned under the next year. The rising curves with time for the treated surface soils are most noticeable for the separate plots in Figures 2 and 3, and also for these as averages in Figure 6. This effect of the clover is marked, first in raising the nitrogen content of the soil above that of the soil at the outset by 346 pounds as an average of the gains as given in Table 4, and second, in bringing it much above the untreated soil in which the nitrogen content was lowered with time. The effect of the clover addition on the nitrogen of the soil must be measured by this difference, which, therefore, makes the increases in nitrogen the more significant. The changes in the total soil nitrogen by the organic matter additions are summarized graphically in Figure 8.

Incorporation Versus Application of Clover on Soil Surface—Where the clover was incorporated regularly when applied, the nitrogen content of the treated soil was raised less over the untreated plot than where this legume was applied on the surface. This indicates a more complete decay of this material and a less amount of its nitrogen remaining in the soil in the former than in the latter condition. The decay processes liberating nitrogen and changing it into a soluble form to be lost by leaching are more active in the former conditions, so that the accumulated nitrogen residue is smaller where the green manure is promptly mixed with the soil. This difference as based on the figures from the three-year advancing averages is indicated by the distance between the two upper graphs in Figure 6 and shown as a separate graph in Figure 7. Calculated from the data in Table 3, this average annual difference is 69.5 pounds, or calculated from the averages of annual analyses, the figure is 60 pounds.

Residual and Mobilized Nitrogen—If the increases by each year over the preceding year as given in Table 3 and Figure 7 are averaged on a single year basis, then 36.5 pounds of nitrogen are added to the nitrogen in the soil each successive year where the organic matter is immediately incorporated, and 40.1 pounds where it is applied on the surface to be incorporated the next year. This represents the residue left in the soil from an application of 106.2 pounds of nitrogen. Hence the difference between 106.2 and 36.5 or 69.7 pounds is the amount that became soluble annually in the former while 66.1 pounds were mobilized in the latter. Thus, 3.6 pounds more, or slightly more than 3 per cent, of the applied nitrogen was made soluble and available for plant use where clover was plowed under immediately than where it was left on the surface and incorporated into the soil a year later.

If one uses the final figure of the three-year advancing averages for the four treated plots, or the average of the two treatments in Table 4 which was 2,582 pounds, and that for the untreated plots which was 1,970 at the close of the treatment, then the clover additions increased the total nitrogen in the soil by 612 pounds during the 15 years, or 40.8 pounds per acre per year. Since 106.2 pounds were applied per acre annually, about 38 per cent of the annual application contributed to the increase of the residual nitrogen in the soil as based on the 15 year period. Consequently, the balance of the application, 62 per cent, or 65.4 pounds per acre, was mineralized by conversion into soluble nitrogen through the decay processes of the soil. We may say then that slightly less than two-thirds of the green manure equivalent was transformed annually into available or leachable nitrogen, while slightly more than one-third was converted into the more stable soil organic matter as an addition to the reserve or stored nitrogen in the soil.

Clover Needed to Maintain the Soil Nitrogen in a Fallow Soil—While the clover treatments of 2 ½ tons per acre annually raised the residual nitrogen in the soil above the untreated soil by 612 pounds during the 15 years, the untreated soil was declining for a total of 115 pounds. Thus, almost one-fifth (18.8 per cent) of this addition would have been needed to maintain the soil at the level present at the outset. In terms of the clover application, this would have required a half ton per acre annually. In the light of this it would seem that a two-ton application once in a four-year rotation would be required to maintain the nitrogen reserve in a fallow soil at the level found in the soil at the beginning of the study.

Nitrogen Levels in the Subsurface Soils—An average of the fluctuations in the nitrogen contents of the subsurface soils as given in Table 2 and graphically in Figures 4 and 5, reveals that under the plots with no treatment there is a slight decline in the nitrogen content, while the differences between the nitrogen contents at the outset and at the close in Table 4 suggests insignificant change. For the treated plots, there are slight subsurface gains which are larger where the organic matter is incorporated into the surface soil than where left on the surface for one year.

Since the subsurface soil is more impervious than the surface, and since the still deeper layers are even more impervious, this suggestion of loss of nitrogen in the untreated and gain in the treated soils indicates its movement downward on this more level and limited area even into and through this layer of heavier texture. These changes, however, are very small and can be taken only as indications. They are of much less dimension than those in the surface soils and seem suggestive mainly because they are in agreement with the direction of the changes in the surface layers.

TABLE 1: Annual nitrogen content of soil with red clover incorporated as compared with its application on the surface each year (pounds per 2 million).																	
Series	Clover Treatment	Soil Stratum	Years														Average all Analyses
			1918	1919	1920	1921	1922	1923	1924	1925	1926	1927	1928	1929	1930	1931	1932	
North	Surface application	Surface	2256	2370	2198	2407	2495	2511	2611	2274	2458	2405	2541	2516	2537	2449	2562	2439
		Subsurface	1824	1680	1617	1823	1803	1711	1718	1613	1844	1627	2003	2020	1955	1734	1769	1782
	No treatment	Surface	2132	2215	1802	2044	2145	1938	2155	2023	2037	1835	2015	1643	1994	1918	1825	2005
		Subsurface	1752	1596	1521	1575	1625	1711	1652	1558	1623	1602	1621	1702	1775	1568	1520	1626
	Incorporated	Surface	2272	2349	2052	2199	2288	2482	2231	2297	2265	2444	2445	2531	2516	2440	2697	2367
		Subsurface	1958	1762	1701	1879	1861	1917	1808	1756	1776	1906	2029	2037	1984	1823	1891	1872
South	Surface application	Surface	2293	1999	2204	2329	2546	2571	2586	2404	2589	2608	2529	2601	2716	2632	2633	2482
		Subsurface	1879	1990	1691	1830	1894	1931	1827	1733	1793	1769	2019	1803	1872	1914	1777	1848
	No treatment	Surface	2276	1943	2148	2145	2239	2195	2214	2005	2185	2214	2029	2055	2107	1987	1992	2115
		Subsurface	1843	1820	1644	1790	1783	1707	1826	1638	1724	1833	1756	1775	1910	1671	1745	1764
	Incorporated	Surface	2269	2371	2205	2359	2569	2304	2529	2263	2424	2527	2568	2452	2665	2589	2552	2443
		Subsurface	1728	1768	1564	1649	2449	1677	1815	1615	1625	1716	1840	1680	1979	1587	1749	1713

TABLE 2: Annual nitrogen content of soil with red clover incorporated as compared with its application on the surface each year with data as advancing three year averages (pounds per 2 million).

Series	Clover Treatment	Soil Stratum	Years												
			1920	1921	1922	1923	1924	1925	1926	1927	1928	1929	1930	1931	1932
North	Surface application	Surface	2274	2325	2366	2471	2539	2465	2447	2379	2468	2487	2531	2500	2516
		Subsurface	1707	1706	1747	1779	1744	1680	1725	1694	1824	1883	2006	1916	1832
	No treatment	Surface	2049	2020	1997	2042	2078	2038	2071	1965	1962	1940	2004	1956	1912
		Subsurface	1623	1564	1573	1637	1662	1640	1611	1594	1615	1641	1699	1681	1621
	Incorporated	Surface	2224	2200	2179	2323	2333	2336	2264	2335	2384	2474	2498	2497	2551
		Subsurface	1807	1780	1813	1885	1862	1827	1778	1810	1901	1990	2016	1948	1899
South	Surface application	Surface	2165	2177	2359	2482	2567	2520	2526	2533	2575	2579	2615	2649	2660
		Subsurface	1853	1837	1805	1885	1884	1830	1784	1765	1860	1863	1898	1863	1854
	No treatment	Surface	2122	2078	2177	2193	2216	2138	2134	2134	2142	2099	2063	2049	2028
		Subsurface	1769	1751	1739	1760	1772	1723	1729	1731	1771	1788	1813	1785	1775
	Incorporated	Surface	2281	2311	2377	2410	2467	2365	2405	2404	2506	2515	2561	2568	2602
		Subsurface	1686	1660	1676	1663	1746	1702	1685	1652	1727	1745	1833	1748	1771

TABLE 3. Nitrogen content of soil with increase due to clover additions as green manure (figures as advancing three year averages of pounds per 2 million).

	Years													Average per Year
	1920	1921	1922	1923	1924	1925	1926	1927	1928	1929	1930	1931	1932	
Surface additions	2219	2251	2362	2476	2553	2492	2486	2456	2521	2533	2573	2574	2588	--
No treatment	2085	2049	2087	2117	2147	2088	2102	2049	2052	2019	2033	2002	1920	--
Increase by surface additions	134	202	275	359	406	404	384	407	469	514	540	572	618	--
Incorporated	2252	2255	2278	2366	2400	2350	2334	2369	2445	2494	2529	2532	2576	--
Increase by incorporation	167	206	191	249	253	262	232	320	393	475	496	530	606	--
Difference surface addition over incorporation	–33	–4	84	110	153	142	152	87	76	39	44	42	12	69.5
Annual increment over preceding year, surface additions	--	68	73	84	47	–2	–20	23	62	45	26	32	46	40.1
Annual increment over preceding year, incorporation	--	39	–15	58	4	9	–30	88	73	82	21	34	76	36.5
Annual decrement, no treatment	--	36	–38	–30	–30	59	–14	53	–3	33	–14	31	32	9.5

TABLE 4: Nitrogen content of soils at the outset and at the close of study with clover incorporated and with its surface application. (Averages of two series of plots).

	Surface Application of Clover	No Treatment	Incorporation of Clover
Surface Soils			
Outset*	2219	2085	2252
Close*	2588	1970	2576
Gain	369	–115	324
Annual Analyses (average)	2461	2060	2405
Subsurface Soils			
Outset*	1780	1696	1746
Close**	1843	1698	1835
Gain	63	2	89
Annual Analyses (average)	1815	1695	1792

*Average of the first three years.
**Average of the last three years.

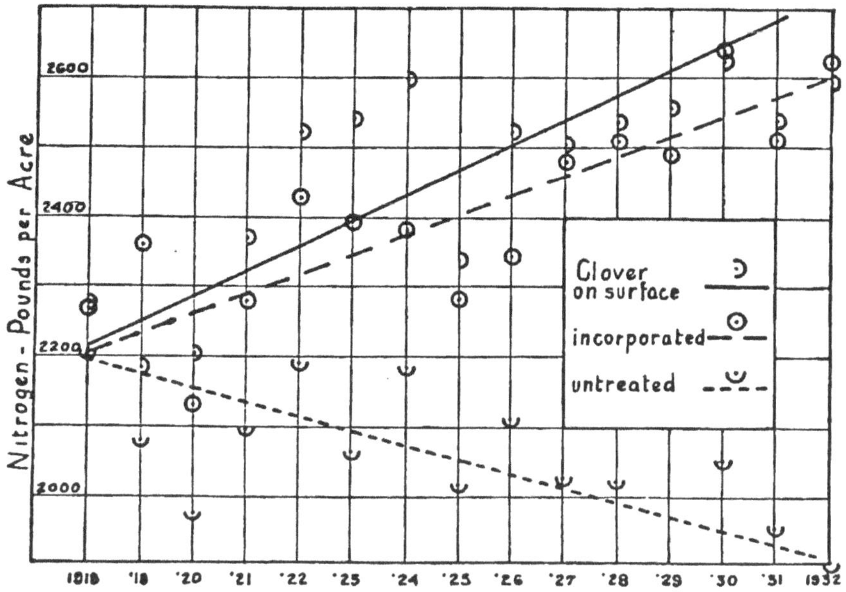

Fig. 1. Annual nitrogen contents of surface soils with red clover incorporated in comparison with its annual application on the surface.

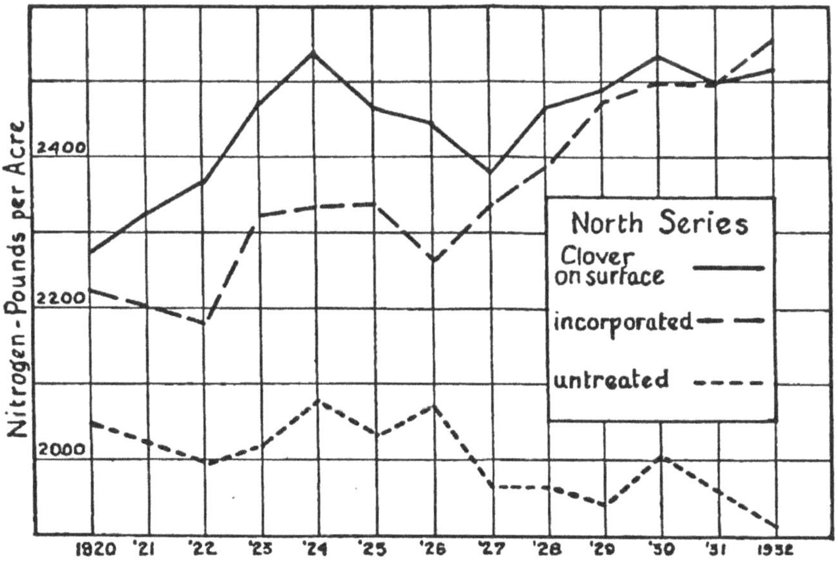

Fig. 2. Nitrogen content of surface soils of north series with red clover incorporated in comparison with its annual application on the surface. (Data as advancing three-year averages.)

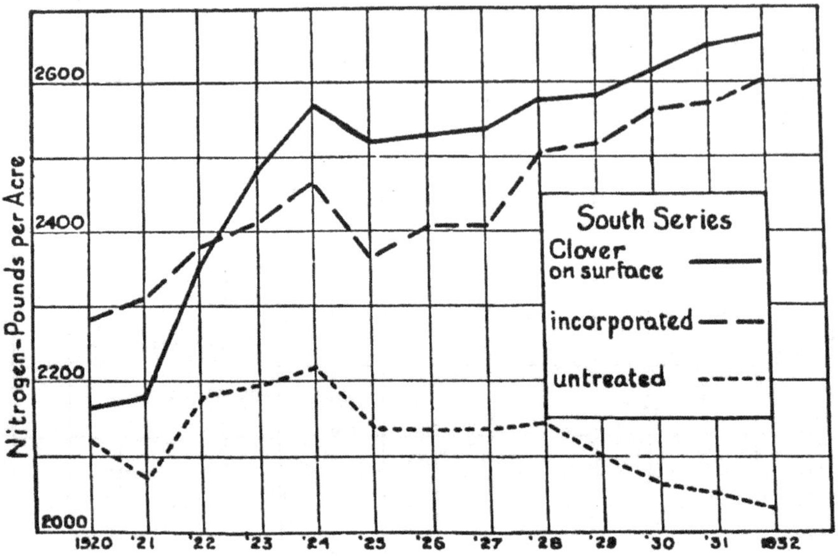

Fig. 3. Nitrogen content of surface soils of south series with red clover incorporated in comparison with its annual application on the surface.

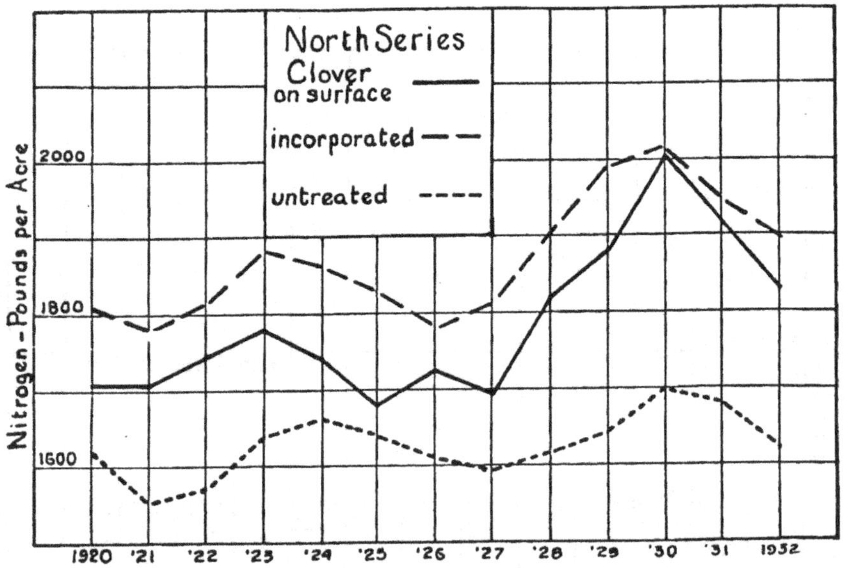

Fig. 4. Nitrogen content of subsurface soils of north series where red clover was incorporated with the surface soil in comparison with its annual application on the surface. (Data as advancing three-year averages.)

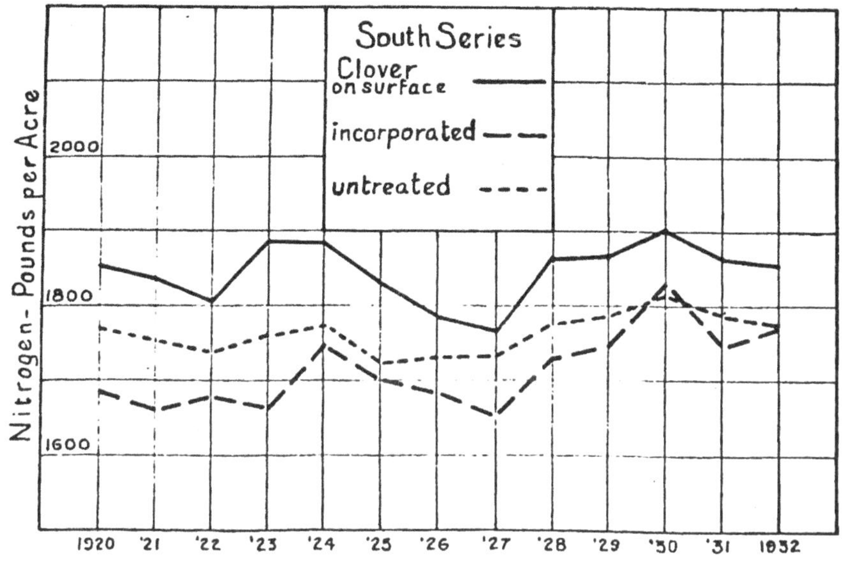

Fig. 5. Nitrogen content of subsurface soils of south series where red clover was incorporated into the surface soil in comparison with its application on the surface. (Data as advancing three-year averages.)

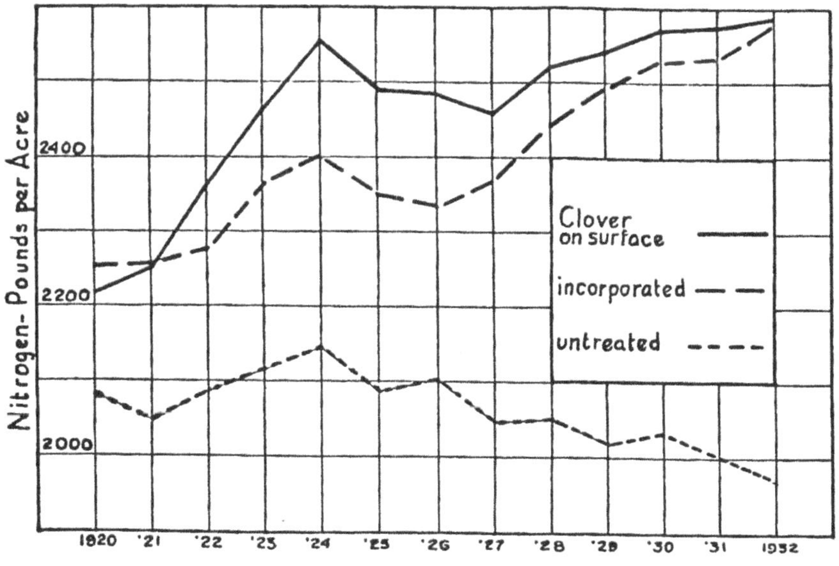

Fig. 6. Average nitrogen content of surface soils of two series of plots with nitrogen incorporated annually into the soil in comparison with its application on the surface. (Data as advancing three-year averages.)

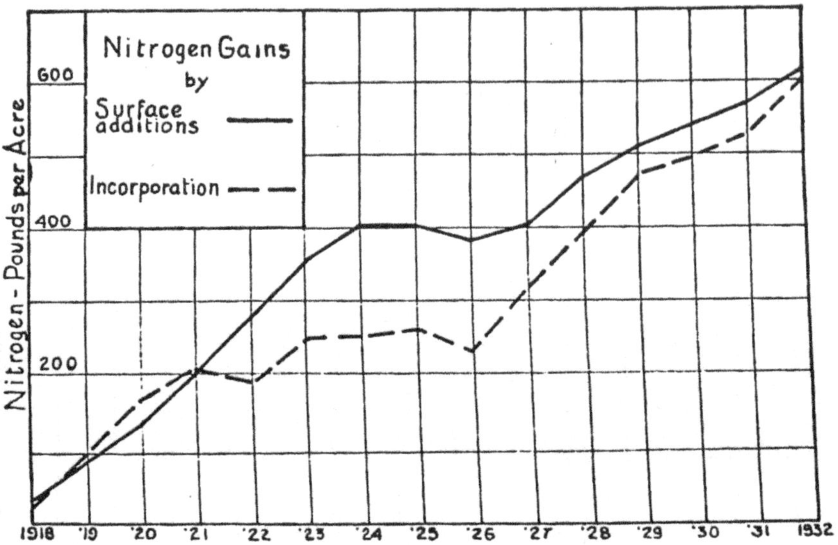

Fig. 7. Gains by treated over untreated plots in nitrogen of surface soils where red clover was incorporated annually in comparison with its application on the surface. (Data from advancing three-year averages.)

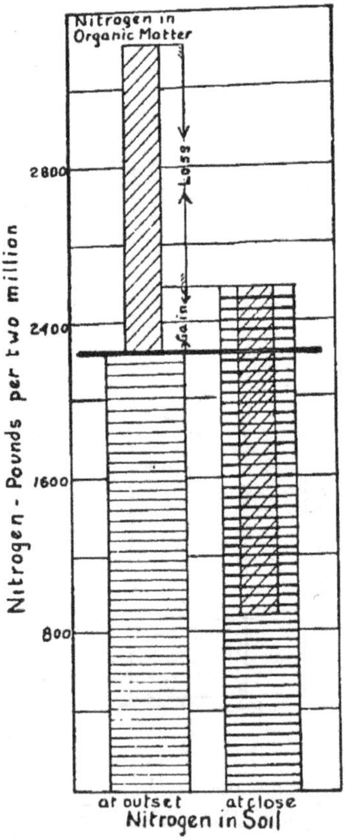

Fig. 8. Nitrogen in the soil at outset plus the total organic matter added in comparison with the total nitrogen in the soil at the close.

CHAPTER 7

Some Soil Factors in Nitrogen Fixation by Legumes[1]

By means of a plant medium of colloidal clay (3) on which various nutrient cations and anions are adsorbed, it is now possible to put the soil phase of nitrogen fixation by legumes under chemical control as accurate as that possible by the water culture methods for plants, or the agar cultures for isolated bacteria. Rather than establish our ideas about nitrogen fixation by piecing together the information gained by each of these performers in isolation, we can now interpret nitrogen fixation by observing the soil, the plant, and the bacteria, all performing cooperatively under accurate chemical control. Though colloidal clay mixed into quartz sand is not truly a soil, yet its simulation of such is so close as to help us evaluate more exactly the significance, in the process of nitrogen fixation, of (a) the soil texture, (b) the pH, or degree of acidity, (c) the degree of calcium saturation of the clay, (d) the fertility in terms of organic matter content, (e) the phosphorus mobility, (f) the exchangeable magnesium content, and (g) the exchangeable potassium content of the soil.

Soils of "Heavy" Texture Give Increased Nitrogen Fixation

It has been observed often that legumes may be growing successfully on a heavy clay soil the degree of acidity of which is such as would cause their failure in most soils. Experiments with colloidal clay media show

[1] *Wm. A. Albrecht, "Some Soil Factors in Nitrogen Fixation by Legumes," Journal Series no. 611 (Missouri Agricultural Experiment Station). Reprinted in Transactions of Third Commission of International Society of Soil Science, New Brunswick, NJ, vol. A (1939), 71–84.*

that increasing amounts of clay at certain degrees of calcium saturation give improved nitrogen fixation and better growth by the legume (1). When electro-dialyzed colloidal clay that was originally at pH 3.6 and was titrated by calcium hydroxide so as to give a range in pH from 4.0 to 6.5 by intervals of 0.5, was added in increasing amounts to a constant quantity of sand, the growth of soybeans showed improvement. Such growth improvement corresponding with more clay, or with a heavier soil texture, occurred for the clays of different acidities when these were pH figures above 4.0. At this latter degree of acidity, very little growth occurred. These results point to the clay fraction as a significant soil factor in the process of nitrogen fixation.

The efficiency with which the soybean plants, for example, can withdraw the calcium from the exchangeable supply in the calcium-saturated colloidal clay was determined. Plants given increasing amounts of such

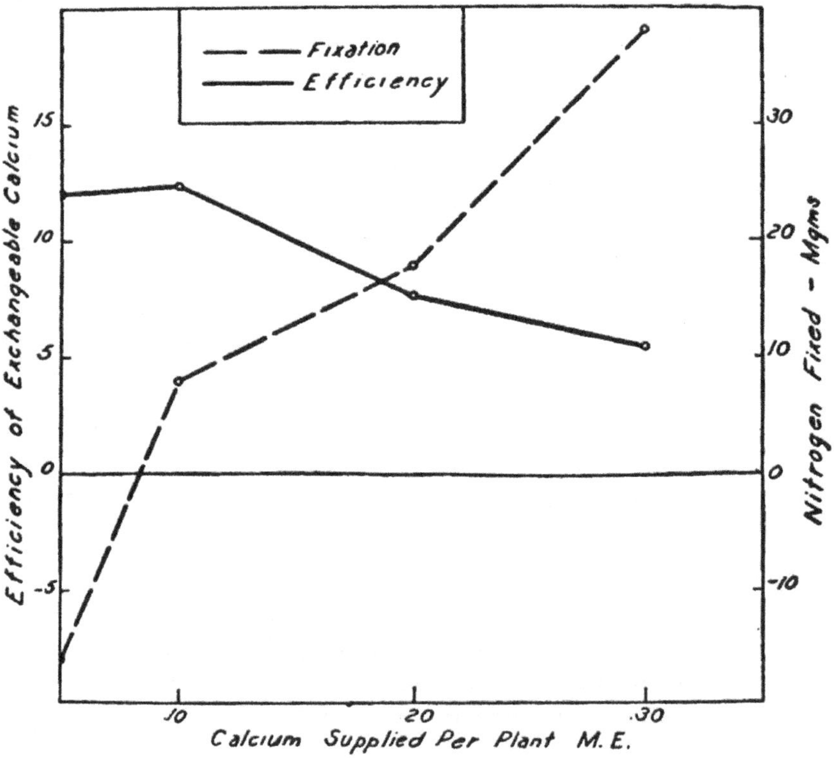

Fig. 1. Nitrogen fixation as related to efficient use of exchangeable calcium with increasing clay content of soil.

clay in sand were analyzed (4), and their nitrogen fixation was determined as related to the efficiency with which they consumed the exchangeable calcium from the clay when they were restricted to other nutrients supplied only in the seed. The results given in figure 1 point out that even with an ample but constant total supply, the maximum efficiency of such calcium use was about 12.5 per cent. This efficiency was lower as the fixation was higher, and as a result, with the most effective nitrogen fixation only about 6.5 per cent of the exchangeable calcium supply was taken, or served the plant.

For these legumes the exchangeable calcium would not be available in toto but only to about one-sixteenth during these 5 to 6 weeks growth. This suggests that before we can use chemical measures of the exchangeable nutrient supplies in the soil as indexes of fertility, it is necessary to determine the efficiency with which these nutrients in exchangeable form serve the different crops.

The Degree of Acidity, or pH of the Soil, and Nitrogen Fixation

The foregoing studies served to emphasize the significance of the degree of soil acidity and the nature of the colloidal clay complex as factors in legume growth and nitrogen fixation. In the first case cited, nitrogen fixation during a 6-week period of growth failed at all pH figures below 5.5, even with increasing amounts of clay in which the total absorbed, or exchangeable, calcium was no greater than 0.2 m.e., or 4 mgm., per plant. Nitrogen fixation occurred, however, and was improved with increasing amounts of clay—and therefore of calcium—at pH figures of 5.5 and higher. With the increasing nitrogen fixation there were larger amounts of calcium in the crop as taken from the clay. These were larger both as the amount of clay increased and as the pH figure increased.

There is an inclination to give emphasis to the relation between the pH figure and the process of nitrogen fixation. The plant growth and this fixation process were influenced more, however, by the amount of clay—and therefore of the calcium offered—than by the pH. In fact, doubling the amount of clay improved the process as much as, or more than, a reduction of 1.0 pH, or by ten times in the hydrogen-ion concentration. Thus, since the nature of the clay is such that an increased degree of acidity, or a lower pH figure, represents less calcium, then, in reality, nitrogen fixation is related to acidity, or pH, only as this represents a decreasing supply of calcium as a plant nutrient. This statement holds for the studies with soybean plants, the isoelectric point of which was at approximately pH 5.5. At degrees of acidity less severe than this, the soil calcium moved into the plant for improved growth and nitrogen fixation. At degrees of acidity more severe than pH 5.5, the plants took no calcium from the soil and fixed no nitrogen. On the contrary, they lost bases, or positive ions, from the seed supply to the clay, and a growth resulted with plants of less calcium content than that originally represented by the seed. Thus, it is true that the pH can be so low that no nitrogen fixation occurs, but the calcium deficiency is the true cause and the pH merely a contemporaneously associated variable. This process cannot be expected unless significant increase in the calcium content of the plant is brought about by the delivery of this element to the plant by the soil.

Degree of Calcium Saturation of the Clay and Nitrogen Fixation

The nature of the colloidal clay as a complex which can absorb cations—possibly anions—and exchange them with other ions, suggests that the influence of increasing H-ion concentration on nitrogen fixation by legumes may be an effect of the reciprocal, namely, the decreasing degree of calcium saturation, on this Beidellite mineral. A test was made on the influence of the degree of calcium saturation on nitrogen fixation by soybeans. Saturation degrees of 40, 60, 75, 87.5, and 97 per cent of the capacity of the clay for calcium were used when the accompanying ions were hydrogen, giving clays of varying acidity, or when they were magnesium, potassium, or barium, all giving purely neutral clays. Amounts of the clays were taken to give constant amounts of exchangeable calcium per plant.

All of the visible plant characters demonstrated the importance of the degree of calcium saturation, but there was no significant influence of the accompanying ions when these were the non-nutrients, namely, hydrogen and barium. Here again the effectiveness with which the calcium was delivered served as a measure of the plant performance as a nitrogen fixer. Increased degree of saturation gave greater nitrogen fixation, irrespective of the accompanying cation. Along with it went a greater percentage use of the total calcium offered, or a larger supply of exchangeable calcium was delivered to the plant, as shown in figure 2.

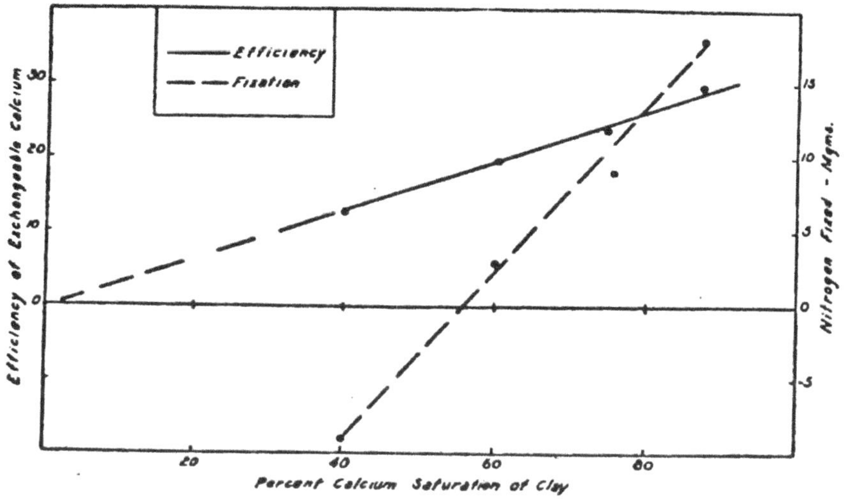

Fig. 2. Nitrogen fixation as related to the efficiency of calcium use at different degrees of calcium saturation of the clay. (H, Ba, Mg, K as reciprocal ions. Mean results of six trials.)

A higher efficiency in the use of the calcium was shown in these trials than in the preceding cases. This efficiency was related to the degree of saturation as illustrated by a straight line in figure 2. This line demonstrated a mean ratio of 0.3215 for the degree of calcium saturation to the percentage of utilization, or a tangent of an angle of 17.8°, for the increased efficiency by calcium with increasing degree of clay saturation by it. It shows forcibly that calcium from the same original total supply is delivered into the plants to a much larger extent when it is on a nearly saturated clay than when on one only partly saturated. As an illustration, the 200 mgm. of calcium distributed on enough clay to saturate it to but 40 per cent, delivered only 21.4 mgm. to the plants. When this same total amount was put on less clay so as to saturate it to 97 per cent, the delivery to the plant amounted to 58.2 mgm., or 2.7 times as much.

As the degree of calcium saturation of the clay becomes lower, more total exchangeable calcium must be present in the soil to be equally as effective as that at the higher degrees of saturation. More exchangeable ions can be provided by supplying more clay. As an illustration, let us suppose that the situation in a silt loam with 15 per cent clay is the same as that for 87.5 per cent calcium saturation in these trials where 29.3 per cent of the total exchangeable calcium moved into the plant. Then let us calculate how much clay equivalent would be required to deliver the same amount of effective calcium, and what quantities of exchangeable calcium would be represented in each of the different degrees of calcium saturation and percentage efficiency cited. Such data are given in Table 1 for the Beidellite clay of which the total exchange capacity was taken as 65 m.e. per 100 gm. of clay.

TABLE 1: Percentages of clay required in a soil at different degrees of saturation to deliver as constant an amount of effective calcium as was found for 87.5 per cent saturation.

	Exchangeable calcium		Effective calcium	Clay required	
Degree of saturation	Per cent used	Pounds per acre	Pounds per acre	Pounds per acre	Clay in soils
Per cent 87.5	29.3	3,412	999.8	300.000	*Per cent* 15.00
75	23.8	4,200	999.8	430.769	21.53
60	19.7	5,075	999.8	650.641	32.53
40	12.3	8,128	999.8	1,563.076	78.15

Since the effectiveness of exchangeable calcium decreases decidedly as the percentage saturation decreases, the amount of clay demanded to provide constant amounts of effective calcium, according to these studies, rises so rapidly that a silt loam at 87.5 per cent calcium saturation is as effective in providing calcium as is a clay soil with 78 per cent clay at 40 per cent saturation. These figures for efficiency of calcium at different degrees may be far from the actual truth in the field, but they serve, nevertheless, to emphasize the fact that calcium delivery to the plant is much more efficient at the higher degrees of calcium saturation of the clay.

These results showed that growth of the plants occurred at all these degrees of saturation by calcium, but nitrogen fixation became significant only at the higher degrees. Thus, plants may be growing but failing to fix nitrogen because of a deficient degree of saturation rather than because of a deficiency in the total supply of this element in exchangeable form. The results suggest that nutrient additions to the soil might well be applied so as to saturate a limited soil volume, with consequently more efficient use by the plants than when applied through a greater soil volume with correspondingly lower degree of soil saturation. This suggests drilling rather than mixing fertilizers with the soil.

Organic Matter Content of the Soil and Nitrogen Fixation by Legumes

Since it is now recognized that the mineral clay complex of the soil unites chemically with residues from organic matter decomposition (5), an attempt was made to simulate a clay having part of its saturation capacity taken by calcium and part by organic matter. This was brought about by giving the clay variable degrees of calcium saturation as previously cited, while the remainder of its capacity was given methylene blue, an organic compound of great molecular size and weight. Soybeans were grown on this medium, delivering the same total exchangeable calcium as that in the preceding test, and their nitrogen fixation was noted. They were compared with a series using hydrogen as the ion accompanying the calcium, with the results given graphically in Figure 3.

The most noticeable result was the fact that the efficiency of calcium use by the legume was not related to the degree of saturation. Rather, the efficiency was almost a constant (48.5–51.0 per cent, mean 49.4 per cent) for all degrees of saturation. Here with the larger organic cation accompanying the calcium, the efficiency of calcium use was more closely related to the total supply of exchangeable calcium than was the case when less strongly adsorbed ions, H, Ba, Mg, or K, accompanied the calcium. Such results suggest that, probably, when a soil is liberally supplied with partly decomposed organic matter, this may exercise beneficial effects on nitrogen fixation indirectly. These may be exercised through the replacement of the hydrogen on the clay by a very large organic molecule, the high adsorption energy of which holds it so firmly that only calcium and other more mobile cations move from the clay into the plants. Such a view may help in the understanding of why certain degrees of acidity in soils of decidedly higher organic matter content are less disturbing to nitrogen fixation by legumes than are these same degrees of acidity in soils almost devoid of organic matter.

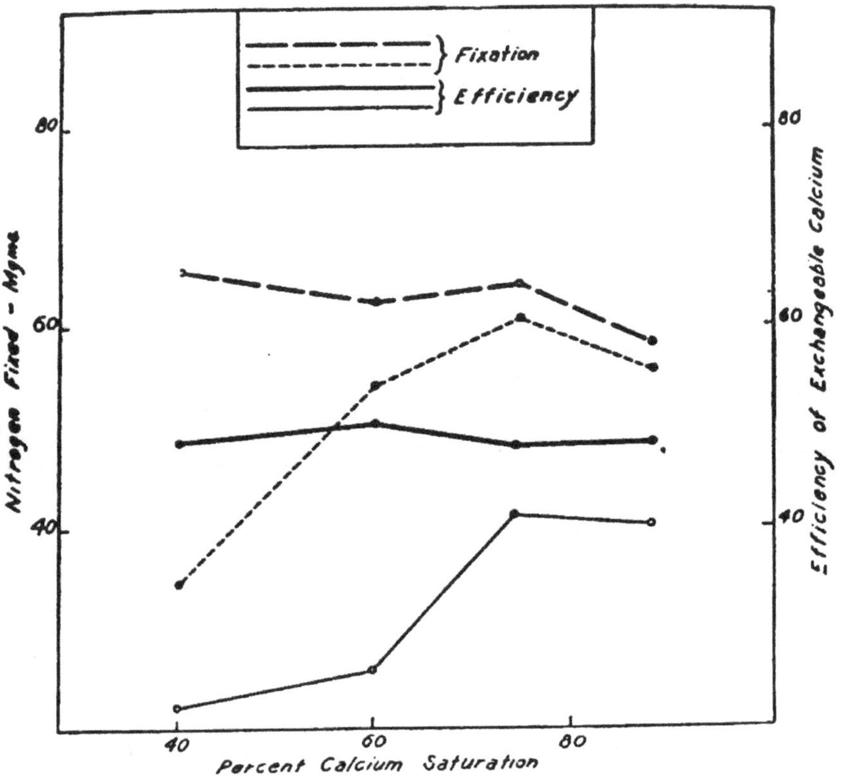

Fig. 3. Nitrogen fixation as related to the degree of calcium saturation when reciprocally accompanied by methylene blue or hydrogen. (Heavy line—methylene blue; light line—hydrogen.)

Phosphorus Mobilization and Nitrogen Fixation

The phosphate anion may be adsorbed by the colloidal clay, hence the clay medium lends itself to the study of this anion, as well as to cations, as a factor in nitrogen fixation by legumes. The adsorption of the phosphorus is influenced by the adsorbed calcium. Studies of the effect of phosphorus on nitrogen fixation by the colloidal clay medium for soybeans show that the mobility of the phosphorus at low levels and its activity in plant growth and nitrogen fixation are linked with the calcium behavior.

Variation in the phosphorus levels of the colloidal clay was without effect on the nitrogen fixation at low levels of calcium. Unless significant amounts of the latter element were present, the plants failed to metabolize into their tissues even the phosphorus present in the seed. Unless the entire phosphorus content of the seed moved into the plants, the latter failed to fix nitrogen during a growth period of 6 weeks. When ample calcium was applied, the quantity of phosphorus in the seed was enough to give significant nitrogen fixation during this period. Four trials were made with increasing amounts of calcium per plant to note the mobilization of the seed phosphorus and its possible relation to the nitrogen fixation. The mean results of the seed phosphorus metabolized in the plants as related to the calcium content of the crop and to the nitrogen fixation activities are shown graphically in Figure 4.

It is interesting to note that with a mean total of 49 mgm. of phosphorus in the seed at the outset, much of this was returned by the plants to the soil unless 0.2 or more m.e. of calcium per plant was present. Further, significant nitrogen fixation occurred only when the total seed phosphorus was moved into the plant and when none was returned to the soil. Thus, there was a close relation between the phosphorus supplied to the plant and nitrogen fixation. These two, in turn, were related to the amount of calcium supplied.

Chemical studies of legumes in the field (2) point to the increased mobilization of phosphorus from the soil into the crop by the calcium treatments. This increase amounted to as much as trebling the phosphorus taken from the soil. It held true for corn, a nonlegume, as well as for lespedeza, a legume.

Thus, if increased phosphorus is required for improved nitrogen fixation, and this increased phosphorus in the plant is impossible without a significant supply of calcium, then the latter improves nitrogen fixation, in part at least, through its mobilization of phosphorus into the plant.

It seems, then, that the three elements nitrogen, phosphorus, and calcium are closely connected with nitrogen fixation by legumes. The first two might readily be related to nitrogen fixation, since both occur as constituents of proteins. That calcium should be associated with protein production not so much as a constituent of the final product but as an agency in its production is not a common belief. Recent relation of this element in the contraction bands of muscular tissue (6) and its persistent correlation with nitrogen fixation in these studies, suggest that this element may play a role in the production, renewal, or other metabolic phases of protein.

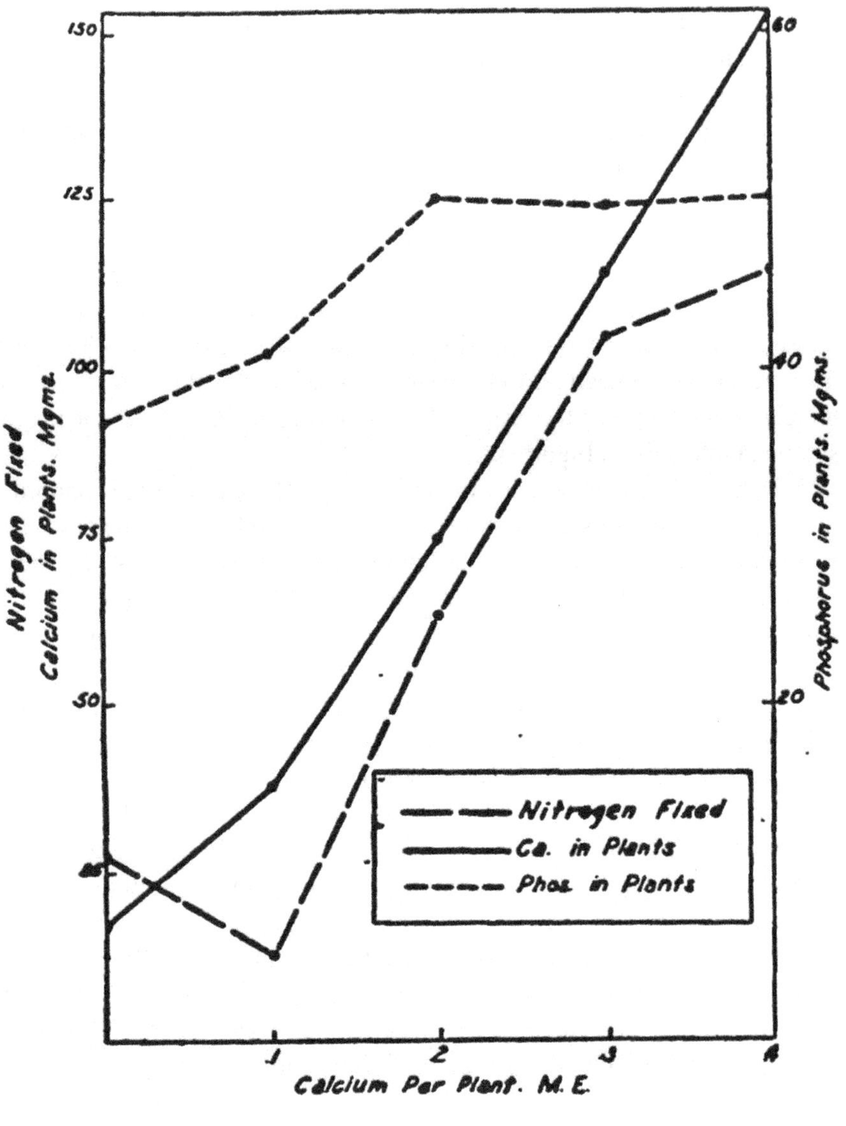

Fig. 4. Nitrogen fixation and seed phosphorus metabolized into soybean plants as related to calcium level provided.

Exchangeable Magnesium and Nitrogen Fixation

When a variable supply of magnesium was used in connection with a constant level of calcium on colloidal clay to determine the significance of magnesium in nitrogen fixation, this element showed a decided influence on the growth and nitrogen fixation by soybeans, as illustrated graphically in Figure 5. That this influence is not directly related to the amounts of magnesium taken by the plants is indicated by the fact that the efficiency with which the magnesium was used is not related to the nitrogen fixation. The magnesium increase, however, gave an increased efficiency in the absorption, by the plants, of calcium from the constant supply of this element at two different levels. This increased efficiency of calcium paralleled the increased nitrogen fixation. The increased total calcium consumption agreed with the increase in plant growth. Such relationships were not evident between the efficiency with which the magnesium was taken and the nitrogen fixation, or between this latter process and the total or percentage of magnesium in the crop. It would seem more logical, then, to believe that magnesium was instrumental in making the calcium effective in the process of nitrogen fixation, rather than that magnesium is directly active in this performance.

That this seems the proper conclusion is indicated by the fact that when magnesium supplemented the calcium, the efficiency of the latter was increased decidedly. The concentration of calcium in the crop was also lower as this element was more effective in nitrogen fixation. Use of atmospheric nitrogen occurred at a calcium concentration in the plant tissue of 0.45 per cent, a concentration lower than those in previous trials, which were commonly 0.50 to 0.55 per cent. At this lower concentration within the plants, in this case using magnesium, the phosphorus and the potassium from the seed were completely metabolized into the plant at the same total amount of calcium per plant required for this in previous trials giving less total growth. Thus, the supplementing of the calcium allotment by magnesium increased the effectiveness of calcium in nitrogen fixation by these legumes. We may, therefore, consider the exchangeable supply of magnesium in the soil as a factor indirectly significant in nitrogen fixation.

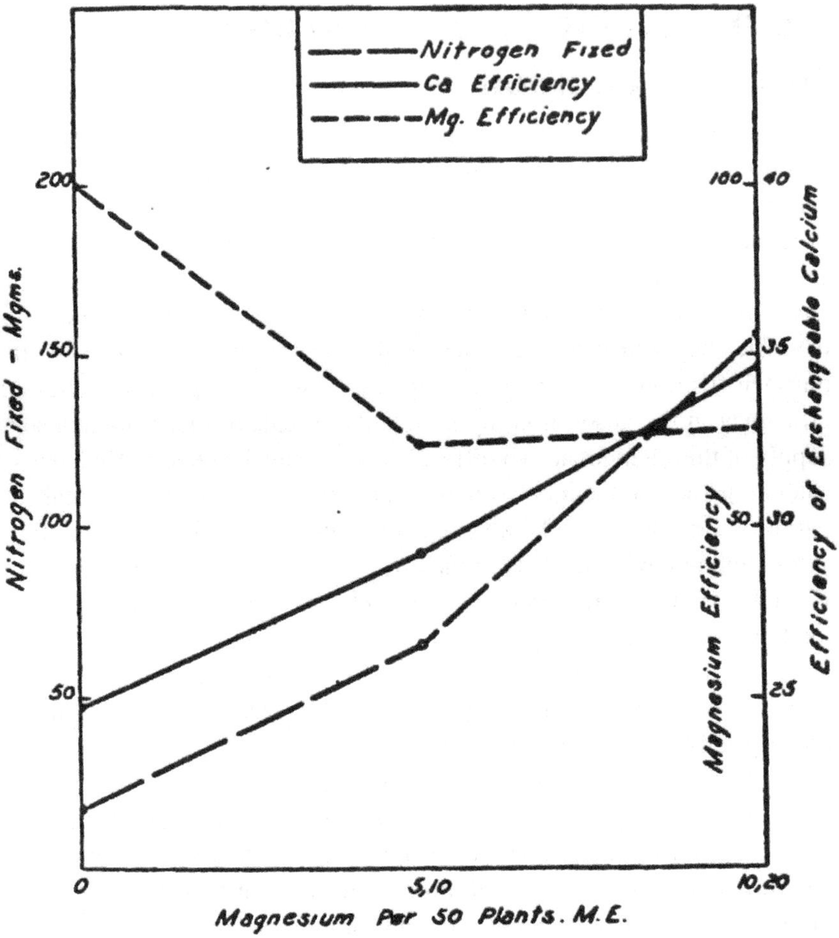

Fig. 5. Nitrogen fixation as influenced by magnesium effects on calcium efficiency.

TABLE 2: Lowered Nitrogen Concentration in soybeans* as related to apparent potassium substitution for calcium.															
Exchangeable cations			Crop											K	Calcium
Mg	Ca	K	weight	Nitrogen		Magnesium		Calcium		Phosphorus		Potassium		Ca	efficiency
m.e.	*m.e.*	*m.e.*	*gm.*	*per cent*	*mgm.*	*per cent*	*mgm.*	*per cent*	*mgm.*	*per cent*	*mgm.*	*per cent*	*mgm.*	*per cent*	*per cent*
5	10	0	14.207	2.86	407	.36	52	.74	105	.25	39	1.01	150	1.36	49.5
5	10	5	14.592	2.56	372	.36	54	.32	46	.18	26	1.90	285	5.93	21.2
5	10	10	17.807	2.19	390	.30	55	.27	48	.14	25	2.15	384	7.96	22.1

* Seed content in mgm.: N =364, Mg = 16.7, Ca = 12.2, P = 39.4, K = 171.

Variable Potassium Supplies and Nitrogen Fixation

When soybeans were grown on colloidal clay with such constant amounts of calcium and magnesium as were previously found effective, but which were supplemented by variable amounts of exchangeable potassium, an increase in growth followed the potassium increments. The nitrogen fixation within the plant decreased, however, with larger amounts of exchangeable potassium. The percentage and the total amount of calcium also decreased decidedly, while the percentage and the total of potassium increased equally noticeably. The data, as given in Table 2 for these elements, show the failure of the plants to use the calcium, whereas their use of the potassium is a distinct feature. These relations and the decreasing concentration of nitrogen are given graphically in Figure 6.

Since the clays were all completely saturated, the introduction of the increasing amounts of potassium along with the constant amounts of calcium and magnesium, lessened the degree of calcium saturation from 66, to 50 and 40 per cent, respectively. That this decreasing saturation is not the sole reason for calcium reduction in the crop is indicated by the use of the calcium at a higher percentage efficiency in these trials than was the case in the previous saturation studies in the absence of magnesium and potassium. It suggests, rather, that the calcium was replaced by potassium. When the calcium concentration is represented by potassium equivalents and these are added to the potassium units as concentration within the plants, the results give almost a constant equivalent for the concentration of these two mineral elements for the three treatments. This suggests a calcium replacement by the potassium. With this replacement came a lowered nitrogen concentration and a lower nitrogen fixing activity.

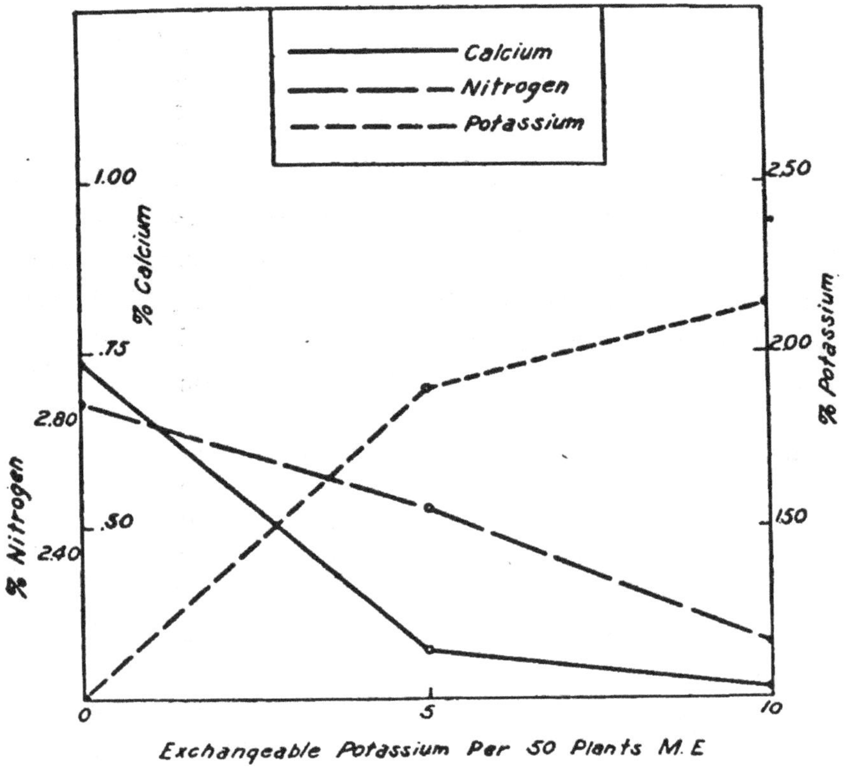

Fig. 6. Reciprocal relation of calcium and potassium concentrations within soybean plants associated with declining nitrogen concentrations as the result of increasing potassium provided.

Possible Calcium-Potassium Ratio in Soil Development Suggested by Nitrogen Fixation Studies

As a theoretical consideration, it may be of interest to note the indication in these results that with a narrow potassium-calcium ratio, the concentration of nitrogen in the soybeans was high, or, as a microbial ration in the form of green manure, these soybeans have also a narrow carbon-nitrogen ratio. With a widening of the potassium-calcium ratio, the nitrogen concentration decreased decidedly, and such plants would represent a bacterial ration with a wider carbon-nitrogen ratio. On soils with the narrow potassium-calcium ratio, green manures would fix more nitrogen to be added to the soil, and this in turn would hold more carbon to build up the soil organic matter. This raises the question whether we might not expect soils in partly developed, or relatively immature, condition with large amounts of exchangeable calcium, and therefore a narrow ratio of potassium-calcium in exchangeable form, to encourage nitrogen fixing plant growths in order to build up the soil organic matter? Conversely then, as the exchangeable potassium-calcium ratio becomes wider, as is true in more highly developed or leached soils, would it be unreasonable to expect the plants to be of a less nitrogenous and a more carbonaceous composition? Such plants are less likely to build up an organic matter reserve in the soil because of the absence of a liberal nitrogen supply where no nitrogen fixation can occur. Chernozem soils, then, with a liberal calcium supply, or possibly a narrow potassium-calcium ratio, probably owe their dark color to their efficiency as nitrogen fixers and carbon retainers. Podzolization, on the contrary, represents calcium removal and consequent reduction in the total calcium to a low level and a wider potassium-calcium ratio to foster nonleguminous plants, even to those of distinctly woody nature. Such reasoning may not be amiss as applied to an ecological series of plants, when one considers that within a single kind of plant, as the soybean, the shift in the potassium-calcium ratio can change the plant from a proteinaceous to a carbonaceous producer.

Summary

By means of studies using electrodialyzed colloidal clay, partly and completely saturated by different nutrient ions, the role of the soil factors in nitrogen fixation by legumes may be more clearly understood. Such studies lead to the following beliefs:

As the soil texture becomes heavier, legumes may be grown more successfully because of the larger supply of exchangeable bases, particularly calcium, offered by the increased clay content, than is at their disposal in soils of lighter texture, or with less clay of the same mineral nature.

The degree of acidity is not of direct significance in nitrogen fixation; but rather its more common reciprocal, the degree of calcium saturation, plays the significant role.

Nitrogen fixation by legumes is closely related to the amount of exchangeable calcium which the crop can take from the soil. The amount so taken is not necessarily related to the total supply: it may be only a small and variable portion of this supply.

The efficiency with which the exchangeable calcium is taken is related to the degree of calcium saturation. The efficiency increases decidedly as the degree of saturation increases. No nitrogen fixation occurs in Beidellite clay, unless the calcium saturation exceeds 50 per cent and the clay carries only adsorbed inorganic ions.

When the larger organic cations are adsorbed on colloidal clay, the accompanying inorganic cations are more effectively used by the plants. This suggests a possible explanation for effective nitrogen fixation by legumes on soils of higher organic matter content, even under significant degrees of acidity.

Calcium plays a possible role in the assimilation into the growing plant of the seed phosphorus, which may even be lost to the soil by soybeans, for example, at low calcium levels. No nitrogen fixation is possible when such loss occurs.

Exchangeable magnesium is of significance in nitrogen fixation possibly indirectly through its influence in increasing the effectiveness of calcium.

Exchangeable potassium of the soil plays a significant role in nitrogen fixation, but as larger amounts are taken in relation to the amount of calcium, the nitrogen fixation in relation to plant growth by legumes is reduced. The potassium may replace the calcium, thus resulting in a widening of the potassium-calcium ratio to the point where possibly the legume does not exercise its nitrogen-fixing ability.

The role of such a potassium-calcium ratio may possibly be helpful in understanding the relation of the organic matter levels in different soils in relation to their degree of calcium depletion, or to their stage in soil development.

References

1. Albrecht, Wm. A. 1938. Physiology of root nodule bacteria in relation to fertility levels of the soil. *Proc. Soil Sci. Soc. Amer.* (1937) 2: 315–327.
2. Albrecht, Wm. A., and Klemme, A. W. 1939. Limestone mobilizes phosphates into Korean Lespedeza. *Jour. Amer. Soc. Agron.* 31: 284–286.
3. Albrecht, Wm. A. and McCalla, T. M. 1938. The colloidal clay fraction of soil as a cultural medium. *Amer. Jour.* Bot. 25: 403–407.
4. Horner, G. M. Relation of the degree of base saturation of a colloidal clay by calcium to growth, nodulation, and composition of soybeans. Missouri Agr. Exp. Sta. Res. Bul. 232.
5. Myers, H. E. 1937. Physio-chemical reactions between organic and inorganic soil colloids as related to aggregate formation. *Soil Sci.* 44: 331–359.
6. Scott, G. H., and Packer, D. M. 1939. Localization of minerals in animal tissues by the electron microscope. *Science* 89: 227–228.

CHAPTER 8

Soil and Nutrition (Via Complete Proteins)[1]

Fertilizers are a help in the better management of our soils for more efficient crop production. However, while we manage soils and crops for larger monetary returns, we must also manage them for better nutrition, better health, and better multiplication of our soil microbes, our plants, our animals and ourselves. We nourish our plants, while aiming for more vegetative bulk or more bushels; we nourish our animals, while hoping for more gain in weight, much of which is water and fat; and we nourish ourselves, with correspondingly the danger of too much fat and degenerative diseases resulting, probably because of the absence of the fuller appreciation of the basic fact that the nutrition of all life forms is dependent on the fertility of the soil.

When food and feed are now accorded increased attention because of the mounting human population, but under declining sheep, pig, and cattle populations combined, why should we not turn our soil-improving efforts via fertilizers, and all else, toward the production also of more proteins? When the production of carbohydrate bulk for filling and fattening has become a more disappointing criterion for our agricultural efforts in production—and maintaining healthy animals, and people—isn't it high time to direct our aims in production toward growing more proteins? Should this not be the aim when these widely deficient food components, especially in their complete array of essential amino acids, represent, (a) growth in terms of cell multiplication (not just waterlogging and fattening); (b) protection against "diseases," that is, against invasions by foreign

[1] *Wm. A. Albrecht, "Soil and Nutrition (Via Complete Proteins)," presented at 30th Annual Convention California Fertilizer Association, Carmel, California, November 9, 1953.*

proteins; and (c) fecund reproduction for maintenance of the species? In our use of fertilizers, then, we must focus attention on making those soil treatments connect the soil more specifically with the most complete nutrition in which the soils and fertilizers can possibly function. Under this as a newer criterion for the services by fertilizers, the industry centered around these commodities will render a larger service for others and itself by assuming the responsibilities of studying nutrition. Nutrition must be studied not only in terms of carbohydrates and calories, but also in its fullest meaning of growth, protection and reproduction for microbes, plants, animals, and man, or the entire biotic pyramid for which the fertility of the soil is the foundation.

Fertilizers and their functions in the soil will be better understood when many basic facts about soils, their chemodynamics in terms of their contents of both secondary and primary nutrient elements, their mineral reserves, the synthetic food services by crops, and the whole ecological pattern of all life in relation to the climatic-fertility pattern of our country, for example, are more clearly comprehended. Fertilizers will be better understood and more wisely used, also, when some erroneous concepts are eliminated, and when practices in soil management can be built on what is truly soil science. Only critical research will weed out the erroneous practices more rapidly and put our own agricultural survival under higher security. Listed and outlined herewith, rather than discussed in details, are some basic facts connecting the nutrition of all forms of life in the biotic pyramid closely with soils, more particularly via the proteins which are so universally deficient and are becoming more costly.

Some basic facts of soil science connecting fertilizers and soils with better nutrition

Our national fertility pattern resulting from our climatic pattern gives corresponding pattern to crops, livestock, nutrition, health and other aspects of all life and puts the soil in control. The varied concentration of farms in the different regions of the United States outlines the varying chemodynamics of the soil for nutrition in a pattern duplicated by that of the radio efficiency; namely, more farms are in the same areas with better radio service. When it is the fertility salts of the soil that are both the soil's conduction of electricity and the soil's production of crops, shall we not see the variable chemodynamics of the elements of soil fertility responsible for both the better crops, the better radio transmission, and better physiological settings for higher forms of life?

The patterns of these activities are determined by the pattern of the climatic forces (rainfall and temperature) as they have been decomposing the soil minerals to keep them growing our feeds and foods of higher protein values for better maintenance of all life. Shall we not connect soil and nutrition more directly as we study and appreciate these larger national patterns?

Plant nutrient elements and compounds adsorbed on and exchangeable from the soil colloids—and not necessarily in aqueous solution—are the fertility "available" to the plant roots. The clays, i.e., the colloidal fraction, of the soil are seat of these chemodynamics. The nutrient elements are adsorbed there from solutions of fertilizers, for example. They are moved from there to the plant roots. Such elements in soluble forms do not move into the plant root along with the water necessarily because the fertilizer salts containing them are soluble, or so-called "available." Water movement into the root follows one set of laws; nutrient movement there follows another set.

When the clay fraction—a negatively charged colloid—is highly stocked with adsorbed fertility elements, then the soil is nearly neutral in reaction, or it is not very acid. More acidity in a soil means merely less fertility per unit of colloid. Increasing acidity is, therefore, simply decreasing fertility. The presence of the hydrogen is not the detriment. Instead, it is a benefit in connection with fertility. It will help the acid clays in their "processing" of the reserve minerals to make them become more "available."

The acidity, or hydrogen, resulting from the respiration of the root, exchanges itself to the clay for other cations, i.e. positively charged ions, of nutrient service to the plants. The acid on the clay breaks down the reserve minerals to move more nutrient ions from there to the clay and from there on to the roots. Acidity, then, is the chemodynamic power within the soil, but of sunshine origin above the soil to keep the fertility assembly line of agricultural production running for full output.

Measuring the exchange capacity and the stock of adsorbed nutrients in it gives more logical concepts of soil fertility and plant nutrition from it. More clay, i.e. more colloids, in our "heavier" soils gives them more adsorption capacity and more exchange capacity. Thereby it gives them more potential productivity. More clay means more fertility even if in company with a high degree of acidity, hence good plant nutrition is offered on "heavy" soils in spite of low pH values. Our "fight" on soil acidity has been a "blind alley." Our use of gadgets to measure acidity and our devotion to that gadgetry has delayed the realization of the failing plant nutrition brought about by the serious fertility deficiencies. Our concern about soil alkalinity will probably find relief when we study it under a concern about fertility shortage for the crop more than a dangerous degree of alkaline reaction.

Organic molecules, as additional cations adsorbed on the clay and as possible carrier of some anions, are a neglected corridor of our soil fertility research. Organic matter suggests itself of major import in the case of the so-called "Alkaline Soils." In the humid regions, soil acidity has fortunately been connected with troubles in growing legumes, though not in terms of interpreting soil as nutrition for those protein producers. Clay must be recognized as the major soil separate on, and around, which the activities of the nutrient ions are centered. Silt can provide the reserve nutrient elements to be liberated by the acid clay. Sand as quartz grains must remain of skeletal service only.

The adsorbed nutrients and their interactions relative to entrance into the plant roots are research areas offering much in better management of plant nutrition. The suite of ions adsorbed on the clay may be (a) the stock of fertility to be measured as amounts exchangeable, and (b) the collected activities of the many ions to be interpreted in terms of each ion in relation to "the company it keeps" rather than to the amounts exchangeable. It is the measurement of the activities of the ions that becomes significant in determining the amounts of nutrients moving into the plant or to be "absorbed" by the plant, as we say it in common language. This fact is suggested by the common emphasis on the many significant "ratios" of the ions "taken" from the soil by plants, and found in the plants according to the chemical analyses of them. The plant is at the mercy of the soil more than vice versa.

Research in soil fertility and plant nutrition needs to consider more than only some inorganic aspects of the soil-root chemodynamics in relation to soil treatments and composition of plants. One-half of the field of soil and nutrition might be considered as neglected in our failure to study the organic compounds which the plant roots may be taking from the soil for nutritional services. To date, we have been studying only the inorganic elements which plants take from the soil. Only the result of "ashing" the plants has been connected with the soil as plant nutrition. We have not considered seriously the plant's necessity to synthesize carbohydrates as truly "plant food" for the plant's struggle to synthesize from those carbohydrates the proteins, i.e. the necessary amino acids, required for the plant's own nutrition if it is to survive. When tryptophane and methionine are amino acids synthesized by plants, this is not merely a philanthropic service by plants to save our costs of purchased protein supplements for fattening livestock. Shall we not fertilize, then, with the production of complete proteins for the nutrition of the crops we grow and study—the protein deficiencies for the crop's nutrition?

Another portion of the study of soil and nutrition, even in the inorganic phase, is neglected. To date the "solubility" of anions in an extracting reagent is our criterion of anionic behaviors in the soil. We cannot yet measure the "activities" of the anions as we measure those of the cations in the soil. When carbonic acid provides the carbonate anion to exchange for other anions, it is necessary that we develop our concept of how this less active anion can exchange for nitrate, chloride, sulfate, phosphate, silicate, and others of negative charge and of more active behaviors in the soil. The concepts of adsorption—exchange of anions held in the soil—still remain to be developed. This half of the inorganic picture of soil and plant nutrition has not yet been painted in detail.

The climatic pattern of the development of the different soils for nutrition in fullest order has too long been viewed only in terms of water for plant use. The climatic pattern of the effects by rainfall in conjunction with those by temperature is more significant in terms of the soil fertility pattern which controls the ecological array of plants than it is in terms of water alone. This soil fertility pattern, that is, the pattern of plant nutrition gives location to either carbohydrate-producing plants only or those producing proteins too, and those more complete in the delivery of all the essential amino acids required for the nutrition of animals and man. It is a climatic pattern in relation to protein production, more than in relation to any other major factor in nutrition. Protein production and the availability of this factor in nutrition is the major ecological factor in any life form.

The ratios of the calcium to potassium and of other elements to each other, control the synthetic services by the crops. The high ratio of calcium to other elements in connection with legume growth is not a matter of calcium as ammunition in the "fight" on soil acidity. Instead, it is a matter of the generous amount of calcium in connection with any of the crops given to the synthesis of much protein, and the more generous amounts required by plants carrying on nitrogen fixation from the atmosphere for good nutrition. According as the climatic forces of the weathering of the soils shift the ratios of the activities of the nutrient elements adsorbed on the clay, so there may be expected a shift in the carbohydrate-protein ratios in the plant's composition. A high degree of calcium saturation of the soil's colloidal complex is a requirement for higher protein concentrations in the crops. In considering the fertility requirements to improve our soils, then, the broader basic climatic pattern giving different degrees of soil development or varying saturations of the colloidal clay as suites of ions is a most helpful guide to the necessary fertilizers for efficient plant nutrition.

Accurate diagnosis of the fertilizer needs of the soils, not just sales, has become a new responsibility of the fertilizer industry. Fertilizer use has become a matter of prescriptions determined by chemical facts about the soil, and about the physiology of the intended crop. It is becoming more and more a problem of providing larger amounts of materials as sustaining fertility as well as amounts of standardized "starter" fertilizers per acre limited so as to prevent salt injury to the seedling plants. Soil tests are becoming more common guides in the decisions as to what fertilizer to use. They are leading the farmer to understand, and to be more concerned about, his soil fertility problems. It is that growing concern of the farmer which has increased the fertilizer sales more recently.

"Fertilize for more proteins by all means" may well be the slogan for all of us. The services by fertilizers will increase, and thereby sales of them will increase too, when soil and nutrition are more closely connected by our knowledge of the neglected trace elements functioning in our soils, our crops, our animals, and ourselves. Enzymes, the catalysts to speed up biochemical reactions, usually consist of three components, namely, some "trace" element, a protein-like part and a vitamin-like unit. That chain of elements and compounds links the soils with the dynamics of creation and maintenance of life very directly. As tools, the magnitude of the mass of the enzymes is not an index of its magnitude of service, hence "trace" amounts become very important in very small amounts. Significance attaches itself to "trace" elements when they are active in the synthesis of proteins, and when their shortages have been connected with shortages of the most commonly deficient amino acids, methionine, and tryptophane and with the nitrogen fixation process by legumes. In terms of the problem of protein production, the trace elements in a larger number become as significant as is the neglected element magnesium in chlorophyll and its synthesis of the carbohydrates. When the major elements, which we recognize in terms of their amounts available in correlation with the magnitude of the products created, are so badly needed as fertilizers, then surely the "trace"

elements as tools, which in trace amounts are controlling the magnitude of production, become even more significant in our soil treatments.

The limited supplies of fertilizer resources demand knowledge and thrift in their use for more proteins, not just food.

Fertilizers, to date, represent about thirty percent of our agricultural production, our creation and not transformation, of new values, and of food values at that. With the recognized inadequacy of potential fertilizer resources for them to assume the complete production of food, it behooves us to learn the chemodynamics of the production still resulting from the fertility resources from the soil's supply. Fertilizers are still only a bit of nutritional supplement, but the finishing supplement, to what the crop plants rustle up for themselves from the soil. It behooves fertilizer producers then to fit their products into the farmer's need for plant nutrition. In connection with that service fertilizer producers, distributors, and salesmen must therefore be better students of soil science in order to appreciate the importance of

(a) the climatic setting according as the degree of soil development in it represents soil under construction or under destruction;

(b) the clay content and clay nature of the soil representing high or low exchange capacities for extensive or limited stocks of fertility or disturbing amounts of hydrogen ions;

(c) the supply of fertility reserves in the silt fractions virgin to the soil, and washed in or blown in on it, but yet to be broken down by the root activities resulting from carbonic acid output and serving as a sustaining fertility over a long time;

(d) the ratios of the activities of the nutrient cations and anions offered by the soil colloidal organo-inorganic complex representing suitable amounts and ratios exchangeable for the root's cation, i.e. hydrogen, and its anion, i.e. bicarbonate;

(e) those ratios controlling the plant's synthesis largely of carbohydrate bulk only, or its synthesis also of proteins complete in all the amino acids required for the nutrition of plants, animals, and man; and

(f) the organic compounds in the soil to supply not only those required for plant nutrition of highest order, but all compounds contributing secondarily, via microbes as well as plants, to make the soil provide nutrition for us as body growth, body protection against degenerative diseases, and body reproduction in healthy offspring.

Fertilizers must move into the high service of real nutrition. The producers, distributors, and salesmen of them have a larger opportunity but a larger responsibility to give that service. The provision of it is a great challenge. We are confident that it will be accepted readily by the men in the fertilizer industry.

CHAPTER 9

Changes in the Nitrogen Content of Stored Soils[1]

Certain determinations for soil nitrogen at the Missouri Agricultural Experiment Station have shown marked evidence of increases having taken place during storage. As the soil samples in question had been stored in containers which were not airtight and as some of the samples were moist, the possibility of bacterial fixation at first suggested itself. As the room where the samples had been stored adjoined the laboratory in which more or less ammonia was being used it seemed more probable that the increase was due to direct ammonia adsorption.

Literature on the adsorption of gases by soils puts ammonia as one of the gases most easily adsorbed. Schlösing found that a moist soil exposed to the air adsorbed nitrogen at the rate of 38 pounds per acre per annum, which was mainly in the form of ammonia. This adsorption took place whether the soils were acid or alkaline, dry or wet (7). A. D. Hall, of Rothamsted (1), after studying ammonia absorption from the air by means of sulfuric acid, states that "the maximum absorption per annum amounts to less than a pound per acre." In testing the amounts of ammonia that soils will adsorb from an unlimited supply, Muntz and Gerard found that 1 kilogram of garden soil took up 5.38 grams of ammonia (4). The presence of various substances in the soil increases the adsorptive capacity for ammonia. It has been shown that calcium carbonate (7), ferric hydroxide, and organic matter (2) have marked effects, while moisture is not as effective as one would suppose (6).

Lyon, Fippin, and Buckman (3) state that—

[1] *Wm. A. Albrecht, "Changes in the Nitrogen Content of Stored Soils," Journal of the American Society of Agronomy 10, no. 2 (February 1918): 83–88.*

Absorption of gases by soils is largely an adsorption phenomenon, the gases being condensed on the surface of the particles. The absorption is greater, the finer the particles of soil, but this increase is not directly proportional to the increase in surface, large particles apparently having a greater absorptive power than their surface would indicate.

As a surface phenomenon the amount adsorbed will be dependent on such factors as partial pressure, temperature, viscosity of the vapor, physical condition, chemical composition, and others. When the gases are once adsorbed, they are maintained with marked tenacity, as is shown by glass, which holds hygroscopic moisture at temperatures as high as 500° C. Sufficient evidence is available to emphasize the fact that the soil is a powerful adsorber of gases. This mass of information suggested the possibility of contamination of stored samples through this means. The following study was undertaken to find out whether contamination by adsorbed ammonia or bacterial activity was responsible for variation of the nitrogen in stored samples.

The plan of storage and treatment of the soil was as follows. Two soils widely different in nitrogen content were collected; one a Shelby silt loam with 2,325 pounds, and the other a Summit clay loam with 7,950 pounds of nitrogen in 2,000,000 pounds of surface soil. Each sample was thoroughly mixed and divided into three parts. On one part the determinations of nitrate, ammonia, and total nitrogen were made as soon as possible after collecting. The second part was put into a room adjoining the general soils laboratory and spread out on a table. Determination of the three forms of nitrogen were made four weeks later. The third portion of the soil was put into bags in the moist condition and stored in a basement room to dry slowly for analysis two months later.

The analytical methods were those commonly used. Nitrates were determined by extracting the oven-dried soil with N/16 hydrochloric acid, making it alkaline, boiling off the ammonia, reducing with Devarda's metal, and distilling. The ammonia was measured by distilling the soil and magnesium oxide with compressed air and steam, while the total nitrogen determination was according to the official method modified for nitrates by use of sodium thiosulfate. The samples were thoroughly mixed each time and sieved through an 80-mesh sieve for total nitrogen determinations and through a 20-mesh sieve in the other cases. All determinations were calculated on a water-free basis. Each sample for total nitrogen was dried in an oven at 107°C for eight hours and then transferred to a flask for digestion. Calculations of pounds per acre were based on each separate water-free sample. For receiving nitrate and ammonia distillations, a N/28 sulfuric acid was used and duplicates checked within 0.2 c.c. for the nitrates but not so closely for ammonia. For total nitrogen analysis a N/14 acid was used, and determinations again checked to 0.2 c.c. All samples except in a few cases were run in quadruplicate.

Table 1 gives the data from the different nitrogen determinations. The figures are averages calculated from four determinations.

TABLE 1: Variation in amounts of ammonia, nitrate, and total nitrogen of a soil stored under different conditions.

Date of determination and place stored.	Silt soil.		Clay soil.	
	Weight of water-free soil.	Nitrogen.	Weight of water-free soil.	Nitrogen.
	gms.	*mgs.*	*gms.*	*mgs.*
Ammonia Nitrogen				
On day sampled	82.60	1.53[a] (l.48–1.58)	71–55	1.50 (1.30–1.73)
Dried 28 days near laboratory	83.28	2.31(2.14–2.50)	80.37	2.66 (2.57–2.72)
Dried 56 days in basement	82.70	1.45 (1.32–1.60)	78.95	1.58 (1.47–1.73)
Nitrate Nitrogen				
On day sampled	61.97	Trace	53.66	Trace
Dried 28 days near laboratory	63.68	Trace	61.45	0.310 (.255–.357)
Dried 56 days in basement	63.24	Trace	60.37	0.340 (.255–.408)
Total Nitrogen				
On day sampled	9.79	10.96 (10.69–11.12)	9.480	37.64 (37.45–37.84)
Dried 28 days near laboratory	9.80	11.32 (11.19–11.49)	[b]9.46	39.32 (39.19–39.48)
Dried 56 days in basement	9.73	11.21 (11.09–11.38)	9.29	37.77 (37.58–37.92)

[a] Figures in parentheses denote limits of variations in figures from which averages were calculated. Variations in hygroscopic moisture determination were less than 6 milligrams for the 10 gram samples.

[b] Three determinations only.

The total nitrogen on the day sampled in the silt soil was 2,237 pounds per acre in 2,000,000 pounds of surface soil, variations 2,183 to 2,263 pounds; in the clay soil, 7,943 pounds (7,901–7,981). In the soils dried 28 days in the laboratory, the total nitrogen was, silt soil, 2,311 pounds (2,284–2,345), and clay soil, 8,315 pounds (8,286–8,350). In the soils dried 56 days in the basement, the total nitrogen was, silt soil, 2,306 pounds (2,280–2,339), and clay soil, 8,133 pounds (8,097–8,164).

The data given for the ammonia nitrogen in Table 1 indicate an increase in this form when the soil was dried near the laboratory, while there was no increase when stored in the basement room. This indicates contamination by gaseous ammonia rather than bacterial action, for if the latter agent had been responsible the sample stored in the basement should have given an increase in ammonia also. Bacteria seemed to be playing no role in a measurable way.

The nitrates remained largely unaffected by storage, even though the clay loam shows less nitrogen as nitrate on the day of sampling than when stored. No explanation for this is offered.

The total nitrogen showed an increase with storage, particularly when the soil was kept near the laboratory. With silt soil this increase was small, but with the clay loam quite significant, though the determinations are a trifle erratic in the latter case. The data as a whole indicate that there is no significant change in the nitrogen by bacteria.

The above results prompted another series of analyses on the silt soil only, to test the possibility of contamination of both moist and dry soils by ammonia. Several portions of a large sample of soil were treated as follows. One part was analyzed for ammonia and total nitrogen as soon after sampling as possible. A second portion was dried in the laboratory where ammonia was used. The third part was dried in a greenhouse located in an orchard away from any ammonia. As soon as the soil in the greenhouse was well dried, analyses were made on some of it, and the remainder was divided into halves, leaving one half in the greenhouse and transferring the other half in the laboratory where it was spread out near an evaporating dish containing about 60 c.c. of ammonium hydroxide. This procedure was followed since no ammonia was being used in the room at that time. The data are given in Table 2. Determinations were made in sets of four and figures given are averages.

TABLE 2: Ammonia and total nitrogen in silt soil as affected by different conditions of storage.

Time of sampling and place stored.	Ammonia nitrogen.		Total nitrogen.		
	Water-free soil.	Nitrogen in sample.	Water-free soil.	Nitrogen in sample.	Nitrogen in 2,000,000 pounds of surface soil.
	Grams.	*Mgs.*	*Grams.*	*Mgs.*	*Pounds.*
Fresh soil	79.73	1.01 [a] (0.86–1.12)	9.6609	11.248 (11.189–11.288)	[b] 2,328 (2,317–2,335)
Dried in greenhouse 10 days.	77.85	1.23 (1.07–1.37)	9.7320	11.275 (11.189–11.388)	2,316 (2,298–2,340)
Dried in laboratory 10 days.	77.91	2.91 (2.73–3.11)	9.7382	11.474 (11.437–11.537)	2,356 (2,328–2,347)
Dried and stored in greenhouse	78.13	1.72 (1.65–1.78)	9.7662	11.409 (11.371–11.468)	[c] 2,337 (2,328–2,347)
Dried in greenhouse and stored in laboratory near ammonia	78.30	60.43 (60.04–60.81)	9.7878	19.041 (18.823–19.114)	3,890 (3,846–3,906)
Dried in greenhouse and stored near ammonia; samples not heated			[d] 9.7878	18.823 (18.678–18.969)	[e] 3,846 (3,816–3,876)

[a] Numbers in parentheses denote range in variations. Hygroscopic moisture determinations varied no more than 6 milligrams in a 10-gram sample.
[b] Three determinations only.
[c] Two determinations only.
[d] Water-free soil figured from sample above.

The data in Table 2 show beyond a doubt that both the moist and dry soils have taken up nitrogen as ammonia when stored in a laboratory in which ammonia fumes are present in considerable amounts. The sample dried in the greenhouse and then exposed in a dry state to ammonia fumes adsorbed enough to give 1,553 pounds per acre increase in the total nitrogen. In the distillable ammonia there was an increase by drying in the laboratory and an unusual increase for the dry soil left near an ammonia container. The increase in ammonia nitrogen over the sample when first collected amounts to 1,517 pounds per acre, corresponding closely to the increase of 1,553 pounds per acre of total nitrogen. Evidently the use of small amounts of ammonia in the laboratory is sufficient to increase markedly the nitrogen in a soil exposed there. That the ammonia is not held by the moisture present in the soil is indicated by the increase of nitrogen when the dry soil was brought from the greenhouse and stored near the ammonia. This soil had but 2.75 percent of moisture, yet took up the equivalent of 1,517 pounds of nitrogen per acre. Ammonia thus taken up is held strongly enough so that heating for eight hours at 107–108°C does not drive it off. This is shown by the figures for those samples analyzed without first heating them in an oven as compared with those heated. That the nitrogen is taken up in the form of ammonia is shown by the fact that the increase of this element is distillable with magnesium oxide. In the form of ammonia, the nitrogen is held, not by moisture, but by adsorption as a purely physical phenomenon.

This brief study indicates that in case of soils stored in or near a laboratory in other than air-tight containers, there is a grave danger of contamination by ammonia, whether the soil is wet or dry. It indicates further that when moist soils are left to dry slowly there is little danger of bacterial action measurably affecting the nitrogen content.

Literature Cited

1. Hall, A. D., and Miller, N. H. J. On the absorption of ammonia from the atmosphere. *In* Jour. Agr. Sci. 4: 56–68. 1911.
2. Hilgard, E. W. Soils, p. 274–275. 1906.
3. Lyon, T. L., Fippin, E. O., and Buckman, H. O. Soils, their properties and management, p. 367. The Macmillan Co., New York. 1916.
4. Muntz and Gerard. *Cited by* A. Hébert, *in* Expt. Sta. Record, 5: 144. 1893.
5. Patten, H. E., and Gallagher, F. E. Absorption of vapors and gases by soils. U.S. Dept. Agr., Bur. Soils Bul. 51, p. 49. 1908.
6. Reichardt and Bloomtritt. Journ. F. prak. Chemic., 98: 167. Cited by Hilgard, E.W. (2), p. 276.
7. Schlössing, T. Comp. Rend. (Paris) 110: 429–499. Abs. in Expt. Sta. Record, 3: 110–112. 1891. *Cited by* King, F. H., The Soil, p. 122. Also cited by Patten and Gallagher (5), p. 30.

CHAPTER 10

The Nitrogen Problem—Its Climatic Aspects[1]

The provision of protein is always a big problem. When we say "Protein," you think immediately of meat, cheese, milk, eggs, fish and other foods closely connected with animal products rather than with plants. Beans, for example, are a plant seed rich in protein. We use many other plant proteins as food. Proteins are not so easily produced on any soil. Their provision via the plants and animals requires that we use lime and other fertilizers on many soils before many legumes will grow. On some other soils these crops grow naturally. In the feeding of animals, the big problem is that of supplying plenty of protein supplement. Calling for such a supplement is a confession that commonly grown feeds are not high enough in protein even for our animals. Would these animals survive in the wild state on such soils? Are we not growing them under a disadvantage when we must reach out to some other soils area and bring in to these less fertile soil areas some concentrated forms of protein? The protein problem is, in reality, one of the fertility of the soil growing it, more than one of buying it.

Since it is the nitrogen in the protein that distinguishes this food constituent from the carbohydrates; and since only the more fertile soils will help the crops to grow some protein rather than produce only carbohydrates, we need to provide nitrogen in the soil if we are to give our plants the means whereby they can supply us and our animals with the protein we need so badly.

[1] *Wm. A. Albrecht, "School of the Soil: How Soils Nourish Plants: The Nitrogen Problem—Its Climatic Aspects," PhilFarmer (First Quarter 1948): 2; Wm. A. Albrecht, "School of the Soil: How Soils Nourish Plants: Wheat is 'Hard' or 'Soft' According to the Pattern of Soil Nitrogen," PhilFarmer (Second Quarter 1948): 2.*

Air Is the Ultimate Source of Nitrogen

The nitrogen problem is perplexing because the air, or the atmosphere, is the ultimate source of the soil nitrogen that is taken by the plants and passed on to man and other higher animals as protein. It is true that there are seventy million pounds of gaseous nitrogen over every acre of land and sea. But the very fact that this elemental, or gaseous, form has accumulated to the extent of constituting 79 per cent of the atmosphere, is testimony that nitrogen is a kind of "lone wolf" among the nutrient elements. It doesn't readily undertake any matrimonial linkage with the others. It can be "fixed," or caused to come out of the air and combine with carbon, calcium, oxygen, hydrogen, or other elements even as a laboratory performance only under high pressures and temperatures. Soil microbes, both those in the nodules of leguminous plants and those living in the soil independently of such a host plant, have the ability to make this seemingly indifferent element take to combinations, or to be "fixed," with other elements. This does not occur, however, except as a list of conditional requirements of the soil are complied with fully. Getting nitrogen out of the atmosphere, into the plants, and into the soil is therefore a difficult, a slow, and a none too common occurrence.

Unfortunately, only a very few of the microbes and plant species can use nitrogen to synthesize the protein compounds making up their body tissues. There is, however, a host of both lower and higher forms that are utterly dependent on this combined or organic form of nitrogen assembled by these few forms and left as parts of their bodies decaying in the soil. It has always been a very serious problem, then, in nature itself to get nitrogen linked into compounds serviceable in life. Microbes are struggling for nitrogen; plants are struggling for it; and, too, animals and man are struggling for it. This is true because nitrogen characterizes the protein, the substance through which alone the stream of life continues to flow. All of this nitrogen must come to us by way of the soil and the vegetation on it from the ultimate source which is the air.

Even though nature had stocked the virgin soils with considerable combined nitrogen in the form of humus or other organic matter before we took it over, we now know that soils differ widely in the amounts and activities of their nitrogen. Then, too, it is no small nitrogen problem of agriculture to keep up the soil's supply by reaching up to the air through the help of legumes, or even the help of the chemical industries fixing nitrogen. When soil nitrogen means proteins like meat, cheese, and similar foods are so costly, there is pronounced necessity for us to face the problem about understanding how nitrogen behaves in nature and how it functions in the many life forms.

Geographic Pattern of the Nitrogen Supply in Our Soils

The farmer's familiarity with the troubles of growing legumes that take nitrogen from the air for addition to the soil, has long emphasized the needs for calcium, phosphorus, potassium, and many other fertility elements in the soil. It is now common knowledge that a soil infertile in other respects than nitrogen is not favorable to legume growth and nitrogen addition to the soil by it. A soil can't build up in nitrogen by starting on near zero fertility. If, then, a soil has been under high rainfall, and especially if this has been accompanied by high temperature, to leach out its lime and other fertility, one cannot expect to grow those types of legumes that are most efficient in fixing nitrogen and in building within themselves the proteins seemingly serving most efficiently as animal feed. Soils developed excessively or those having pushed their original rock materials nearer to the sea and solution, are naturally not the soils rich in nitrogen.

Such soils have profiles with distinct surface soil horizons characterized by a kind of organic matter that decays slowly. Its nitrogen may be relatively high in total, but is not very active in terms of changing over to the forms like nitrates that are of better use to plants. This is the kind of organic matter requiring tillage to fan the microbial fires for its destruction and for the liberation of its mineral contents. Such soils require lime and other fertilization before they will grow a legume for improvement of themselves in their nitrogen supply. The organic matter grown on such virgin soils was originally woody because the soil fertility was so low as to prohibit vegetation containing proteins. It was an unbalanced ration for the microbes because there was an excess of carbon in the soil organic matter in relation to its nitrogen. The nutritive ratio was too wide even for the bacterial life. Even these lowly life forms causing organic matter decay do not grow well unless given supplements of nitrogen and general fertilizers. They, too, have a nitrogen problem, and seemingly pass it on to the plants and on to the animals and humans. If the virgin soils grew only wood or forests, there was a nitrogen problem also for the virgin vegetation and virgin wild life. Forests are crops giving mainly wood production and not much protein production. Woodiness is the output that is characteristic of the less fertile soils. Protein production is therefore a serious problem on such more weathered soils under higher rainfall and where virgin forests originally prevailed.

The eastern parts of the United States have rainfalls high enough to give highly leached soils and therefore they grew mainly forests originally. Growing proteins has been a correspondingly difficult problem when the forest areas were cleared and put under cultivation. For this reason the by-products of the "hard," or high-protein, wheat in the form of shorts and bran were moved from the wheat belt of the mid-continent, or prairie areas, so readily eastward as purchased protein supplements to the feeds grown on the leached soils there.

In the central states with their lower rainfall this high protein-producing wheat was growing without lime or other fertilizer applications on the soil. Hard wheat was "natural" on the prairies. It was high in nitrogen or in protein because there was ample active nitrogen available in the soil. This had been put there because legumes grew naturally and numerously on the prairies. They grew well and took gaseous nitrogen freely from the air because the soils were fertile with the mineral elements required in their nutrition to make nitrogen fixation possible.

These prairie soils were still highly fertile in the essential mineral respects because the low rainfall that permitted the intermittently growing grass, but prohibited trees, had not leached this fertility out of the soil. Protein production by the virgin prairie herbages built the nitrogen-rich soil with its great depth and its black color resulting from the ample stock of humus. It built the massive buffaloes of the past. Protein production in the domestic crops has been giving high protein output in the forages and grains to make the choice proteins in beef and mutton, to say nothing of high protein in wheat as human food. A careful study of the soils, the crops, and animal production of the United States shows clearly that the protein production takes its pattern according to the fertility pattern of the soil, which is in turn developed according to the pattern of the climate.

Wheat Is "Hard" or "Soft" According to the Pattern of Soil Nitrogen

One need not include all the different crops or kinds of vegetation, nor need one make a traverse across the entire United States to appreciate the problem of nitrogen in its differing degrees on the different soils. One can take the wheat crop in its limited area to see the variation in amounts and function of nitrogen as the protein characterizing element. This crop is a "soft" wheat, or a starchy grain, in the eastern or higher rainfall areas of its production. It is a "hard" wheat, or a protein-rich grain, in the western or lower rainfall areas of its more extensive growth. Kansas illustrates these facts very well.

One might be inclined to reason that the varying rainfall is the cause of these differences in protein content, but such reasoning does not explain the differences in plant nutrition which makes the wheat in eastern Kansas contain 10 per cent of protein while that in western Kansas contains 18 per cent of this nitrogenous food compound. Nor can rainfall scarcely be the controlling factor when the same wheat variety is a "hard" wheat in western Kansas and a "soft" wheat in Missouri, with a variation so much as 80 per cent in protein concentration. Some other factor than water alone must be operative in these variable physiological functions of the nitrogen in the wheat plant.

That the nitrogen supply in the soil and the time of its delivery generously to the wheat crop are most probably in control of the protein concentration of the wheat grain was demonstrated by some studies in the Department of Soils of the University of Missouri. Different series of plots seeded to wheat were given nitrate nitrogen as fertilizer applications at successively later dates by fortnightly intervals beginning with the fourth of April. Both the yields as bushels of grain per acre and the protein concentration within the grain were measured. From these the protein production per acre was calculated.

The "hardness" or concentration of protein in the wheat grain from these different plots increased accordingly as this soluble nitrogen was applied later in the growing season. In some of the wheat, given nitrogen early, the protein was as low as 8.7 per cent, while in some given this nutrient much later, it was as high as 17 per cent. Here was an increase in hardness or protein concentration amounting to almost one hundred per cent when there was the same original soil, the same seeding, tillage and harvesting manipulations, and the same rainfall on all the plots.

The yields as bushels of grain per acre showed the reverse effect. These decreased as the date of nitrogen application in the spring season was delayed. As a consequence, the production of total pounds of protein per acre also decreased.

In the deeper but drier prairie soils the nitrogen is known to be speedily made soluble and to move downward rapidly. There is the suggestion from these experimental facts, then, that very probably this location of the major share of the nitrogen at greater depth in the soil means the use of it by the wheat crop at later periods in the plants' growth because of the later arrival of the wheat roots into those greater depths of soil. Seemingly this arrival of the roots in, and their use of, this larger supply of available nitrogen by them is more nearly timed for the later growth stage of the wheat plant. The use of the nitrogen is later according, then, as the rainfall is less and as the more effective moisture supply and the soluble nitrogen are at greater depths in the soil profile.

This is quite the opposite of the situation prevailing in soils under higher rainfalls. They have shallow surface horizons and their nitrogen supply is limited to this upper layer where the plant roots seemingly find and take most all of their supply of nitrogen early in the growing period of the plants. Here is the basis for a big early vegetative growth and the use of the nitrogen early for it. This is then also the basis for the limited protein elaboration and low concentration in the seed which develops only during the plant's later growth stages. Such timing of delivery of nitrogen suggests that under high rainfall there is much starch production but little conversion of that carbohydrate into proteins.

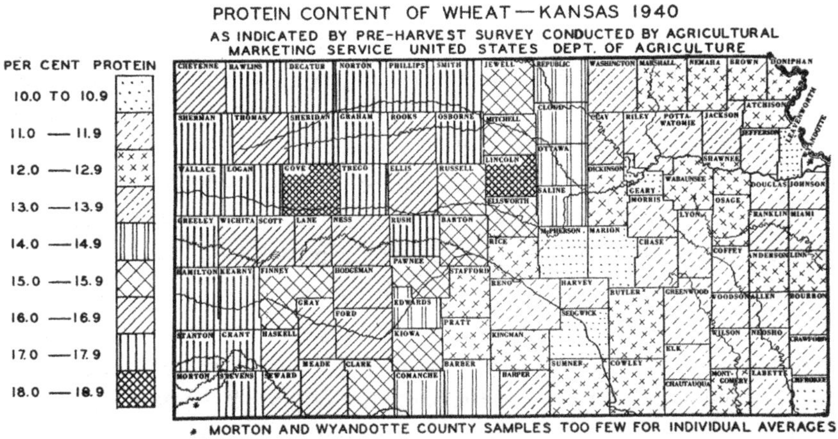

Nitrogen in the soil controls the "hard wheat." Hard wheat has a higher percentage of protein as one goes west across Kansas from her higher to lower annual rainfalls.

Conservation of Soil Means Conservation of Nitrogen

As a pioneer country our westward march meant much soil exploitation. Hard wheat in the Geneseo River Valley of New York once made Rochester the *flour* city in that state. Today it is the *flower* city. Hard wheat has moved westward because the depletion of the nitrogen and other fertility of the soil was pushing it in that direction. We have come to realize that protein production and our harvest of it as meat animals marched along with the hard wheat for the same reason.

In the wake of this westward movement of our protein supplies, the nitrogen problem suddenly stands out boldly. The soil's supply of nitrogen must be increased if protein is to be synthesized according to needs, regardless of whether we harvest it as wheat or as meat. The shrinking chances for us to get beefsteak are sending our minds trailing it back to its source which is the nitrogen in the soil. The higher costs of meat are bringing us to realize that this very desirable form of protein originates not entirely because of the unusual enthusiasm by some farmers as cattle admirers and producers, but because of mineral-rich and consequently nitrogen-rich soils of limited areas making protein-rich forages where the particular combination of climatic forces made such possible. Such handicaps to our appetites turn our minds toward conservation of the soil in a much more real sense than does mere propaganda. Now that our food problem is no longer limited to the United States and that we are assuming a share in the responsibility for it on a global scale, not only calories but also proteins must be grown from the soil. This international problem has compelled us to see that the protein problem anywhere is after all the nitrogen problem of the soil.

CHAPTER 11

Soil Fertility and Soybean Production[1]

What place will the soybean finally find in farm rotations and what will be the long-range effect of this crop on soil fertility? Dr. Albrecht says the effect can be highly beneficial if the meal from the beans is fed back on the farm. The author is chairman of the department of soils, Missouri College of Agriculture. From a paper presented at a soybean dinner in connection with the erection of the Missouri Farmers Association soybean mill at Mexico, Mo. A notable contribution on the subject and "must" reading for every soybeaner.

The days of the pioneers are not all hidden away in past history. They are in the present. These pioneers have dared to demonstrate that if soybeans as food can contribute to victory, they, as farmers were ready to do their part in attempting to grow soybean seed in spite of the once-prevalent contention that the soybean plant is not a profitable seed producer and is suited only as a forage crop in Missouri. This contention is being pushed farther into past history.

Soybean production calls on the soil fertility as help to manufacture substances which livestock gladly eat. It is in terms of this demand for extra fertility that the soybean plant is much different from a weed. It is by providing extra soil fertility that we make a soil healthy enough to grow not only the soybean forage that may be mainly of woody structure similar to weeds, but that we undergird the plant with those helps through which it manufactures during the growing season within its woody frame the

[1] *Wm. A. Albrecht, "Soil Fertility and Soybean Production," Soybean Digest (February 1944): 6–7.*

compounds that are finally mobilized and assembled in the pods as seeds. Seed production as many pods and many seeds per pod, demands healthy soils that are delivering plant nutrients liberally.

Chemical Composition

The chemical composition of the seed is not so variable. That of the forage may be highly so. In this fact there is danger of deception, when forage only is considered. Occasionally plants grow to good size and produce no seed. Again only a small plant may be well loaded with its fruit. Forage is made of carbon, hydrogen and oxygen supplied by the air and water. Seeds demand phosphorus, calcium, magnesium, potassium, and other elements of soil fertility.

When crops fail to produce satisfactory seed yields we are prone to say as was said of the soybean crop when it was first introduced about the time of the first World War, namely, "It doesn't yield well as seeds but it makes much hay." We are prone to forget that if the plant of large tonnage yield as forage fails to make much seed, there can be little soil fertility as minerals and little feed value in the forage as hay.

When the fertility of the soil goes down, then the forage crop truly must "go up in the air." The hay must be just so much fresh air, water, and sunshine with values mainly as fuel or as so much wood. When the crop can't find liberal supplies of soil fertility from which to make itself complete, it can build only a woody framework by means of more air and sunshine. In no unreal sense it is the declining soil fertility that has put the farming business "up in the air" in more ways than one.

We cannot escape our responsibility to the soil by shifting from soybeans as seed crops that are not giving good seed yields per acre to the use of this crop as forage in the belief that the hay will have high feeding value because the plant's pedigree says it is a legume. Regardless of the pedigree, both the forage value of the crop as animal feed, and the seed yield in bushels per acre depend on the soil fertility. Unless the fertility of the soil is mobilized generously the seed yield cannot be high, nor the forage highly nutritious.

To accept much tonnage per acre as roughage in the form of hay when the same crop would be too low as a seed producer is merely to pass the deception on to the animals that eat the hay. Animals so deceived can do none other than pass the deception back to us in the breeding deficiencies, diseases, poor returns, and other troubles commonly passed off as "bad luck." We are witnessing a rejuvenation of the belief that agriculture can survive and can profit only when we keep faith with the land, and when we do not deceive the animals and much less deceive ourselves under neglect of the soil fertility.

In the minds of some persons, soybeans are considered to be an "acid-tolerant" legume, but on soils naturally highly acid, it cannot be true. Because when the coming in of acidity means the going out of plant nourishment, then the belief in an "acid-tolerant" crop is equivalent to the belief that the crop is tolerant to starvation. Plant breeding for acid tolerance would be the equivalent of raising the hope of maintaining life without nourishing it.

We know full well that the soybean plant does not produce seed well on very acid soils. This plant has illustrated its response to calcium fertilization, as the liming results so well show. It has also been demonstrated that soil acidity is no disturbance to this crop when nourishment is provided for it in the form of lime and fertilizers in the lower depths of the surface soil. Where the fertility was delivered to the plants, the crop in turn delivered the good yields of seed beans.

The reduced yield of soybeans on the untreated parts of the fields in contrast to the much better returns where lime and fertilizers were used, suggest that the untreated soils are slowing down in their mobilization of plant nourishment from the soil minerals to the clay and from there to the plant roots. When slightly more than 10 bushels of soybean seed has been the average yield of the state of Missouri, there is good warning that the soils are slowing down so much as to be about to take a rest on their own accord from delivering nourishment to the crops.

A Cooperative Venture

It is particularly fitting on the more level soils, with their significant amounts of clay in the subsoil that there should be initiated one of the first demonstrations of maintaining and rotating the fertility of the soil with the minimum of it lost from the farms. If the soybeans are processed here and only the oil shipped out, can this soybean venture represent any loss of soil fertility from these farms? What is soybean oil but carbon, hydrogen and some oxygen, all coming from the air and water? Sending these air-given elements to market is no drain on the soil fertility of the community. It is dispatching to distant places only so much good Missouri weather. This any one may well, and generously, share with others.

As the oil cake goes from the mill back to the farms to feed livestock, there goes back in it the protein nitrogen taken from the soil. There goes also much extra nitrogen that this legume plant snatched from the unlimited supply in the atmosphere through the help of nodule bacteria as inoculation on the seed. Back to the farm with the press cake will go the extra fertility put on the soil as calcium in the lime, and phosphorus and potassium in the fertilizers to help this legume in running nitrogen-fixing factories on these farms. These fertility elements will go through the animals to supplement the carbohydrate crops as corn and hays fed with it in a balanced ration. The fertility of the corn and hay, too, will be left in a larger measure on the farm.

Here there is being set in motion a community performance of cooperation that should do several things. It will move more purchased mineral fertility into the soil. It will encourage a soil-building program that can deepen the surface soil by building it down into the subsoil. Finally, it will reflect itself in the better reproduction and growth of the animals to say nothing of the sounder health of the people and many other human gains not measured in monetary units.

Here shall be demonstrated a real cooperative venture in soil fertility conservation for community benefit. These soils with their plants, in the animals, and in the people, will do more than bring dollars into the banks and business into the city. They will minister to human health, to contentment, and to the happiness of the community as a whole. Our congratulations go to these pioneers and also our good wishes for their continued success in the constructive venture they have launched in Missouri's northeast prairie section.

CHAPTER 12

Nitrogen Fixation, Composition and Growth of Soybeans in Relation to Variable Amounts of Potassium and Calcium[1]

Introduction

That potassium is absolutely essential for plant growth is well recognized. Although its exact role is not known, it is fairly well established that potassium is necessary in the formation of carbohydrates, and possibly in the synthesis of proteins. The high potassium content of certain species of plants was recognized long ago by the early use of wood ashes as a potash fertilizer. The large, but variable, amounts of potassium contained in plants are in decided contrast to the much lesser potassium contents of animal tissues. The value of manure is partially due to the potassium excreted by the animals, whose metabolic processes are unable to utilize the large amount of potassium contained in the feed.

Any farming system which utilizes manure and crop residues does much to maintain a satisfactory level of potassium in the soil. The production of market crops, usually accompanied by the wasting of crop residues and more recently the limited plowing as a means of erosion control, have led in many cases to shortages of soil potassium for crops. As the available supply of this nutrient element declines, symptoms of an unbalanced plant nutrition become evident, particularly when lime and phosphate are applied.

[1] *Herbert E. Hampton and W. A. Albrecht. "Nitrogen Fixation, Composition and Growth of Soybeans in Relation to Variable Amounts of Potassium and Calcium." Research Bulletin 381 (Missouri Agricultural Experiment Station, April 1944).*

The need for calcium in the nutrition of higher plants has been definitely established. As in the case of potassium, the amount of calcium in the plant varies greatly with the species and with the environmental conditions. Among other things, calcium is unquestionably identified with nodule formation and nitrogen fixation.

Inasmuch as legumes are grown chiefly for their ability to fix atmospheric nitrogen and to produce high quality proteins, wherein potassium and calcium play such significant roles, further carefully controlled studies are needed for a full understanding of the interrelated influence of these two nutrient cations.

Since the colloidal clay technique used so successfully in previous studies of nitrogen fixation permits wide variation in several of the nutrient elements, it was used in this study of the influence on plant growth and composition by variable supplies of exchangeable calcium and potassium on the clay when those of other nutrients were constant.

Historical

Potassium has long been considered one of the essential plant nutrients. Numerous field trials have shown that potash fertilizers are necessary on many soils. However, potassium has been applied chiefly to non-legumes, while lime and phosphorus are usually recommended for legumes.

The ability of the plant to absorb large amounts of potassium has been shown by many studies. Among others, Hoagland and Martin (20) and Loehwing (26) found that the potassium content of the plant reflects the capacity of the soil to supply potassium. The growths of plants are closely associated with their consumption of potassium.

The specific role of potassium in plant nutrition has remained somewhat obscure, although it has been the subject of numerous investigations. From the behavior of oats plants, Dickson (12) concluded that potassium influences meristematic activity. The high degree of mobility of potassium and its more intense absorption during early stages of growth have been demonstrated (22). That increments of potassium bring about increased nitrogen fixation in soybeans has been shown by Ferguson (14). Experiments designed to study the effect of potassium on carbohydrate metabolism have led to somewhat contradictory conclusions. Many investigators (10, 25, 32, 36) are of the opinion that both the formation and translocation of carbohydrates require potassium. The results by Nightingale and his associates (30) caused them to conclude that although potassium is directly or indirectly essential for carbon dioxide assimilation, the translocation of sugars and starch occurs in plants low in potassium. In their work with cowpeas, Janssen and Bartholomew (22) found a close relationship between the concentration of potassium and the total weight of starch and sugars in the plants.

Loew (27) assigned to potassium an essential function in the synthesis of proteins, a role suggested earlier by Stocklasa (33). Loehwing (26) reported that a high ratio of potassium to calcium in plants was associated with a smaller amount of nitrate nitrogen and more organic nitrogen. He concluded that the potassium is an essential factor in protein formation. Similar views were held by Janssen and Bartholomew (22), and Wall (35).

Several workers have reported the beneficial effects of calcium on the growth and nodulation of legumes. The ability of the plant to fix nitrogen is increased as the amount of calcium absorbed by the plant increases (1). Albrecht and Davis (4) concluded that the presence of calcium within the plant is necessary for nitrogen fixation. Horner (21) showed that the amount of calcium taken by the plant depends upon the degree of saturation of the clay with calcium as well as on the level of available calcium in the medium.

A number of writers have suggested the apparent interrelationships which appear to exist between the various nutrient ions. Graham discussed the relative absorption of calcium in relation to magnesium and the significance of the latter in nitrogen fixation by soybeans (17). Evidence of a potassium-calcium relationship has been reported frequently. Fonder (15) found that a physiological balance appears to exist in alfalfa plants between calcium and potassium. A greater calcium content was accompanied by a decrease in potassium. Albrecht (2) suggested that the absorption of larger amounts of potassium by plants in relation to the quantity of calcium may reduce nitrogen fixation in proportion to growth. Potassium may replace calcium to the extent that the widened potassium to calcium ratio reduces the nitrogen-fixing ability of the plant. Similar conclusions have been reached by Hoagland and Martin (20) and by Loehwing (26).

Working with colloidal clay cultures, Ferguson (14) reported an increased consumption of calcium with increments of potassium. Most studies in this connection have utilized aqueous or sand cultures. Needless to say, the conditions existing in colloidal clay cultures should more nearly approximate those of ordinary soils. The relative proportion of nutrient ions absorbed by plant roots may depend largely upon the type of culture in which the roots are functioning. Root absorption from the ionic atmosphere of colloidal particles by exchange (23) may be expected to differ considerably from root absorption from non-colloidal systems. Certainly the energy by which ions are adsorbed by the colloidal soil particles, as well as their relative amounts in the media, should be a factor in regulating their entrance into plants.

The importance of calcium in the fixation of nitrogen by plants has been emphasized. The role of potassium in carbohydrate formation has been rather clearly demonstrated, but the influence of potassium on nitrogen fixation and nitrogen metabolism in legumes as interrelated with those by calcium is still obscure. Our knowledge of the relative consumption of potassium and calcium and their comparative effects on nitrogen fixation and the composition of nodulated legumes is particularly incomplete. It is usually assumed that legumes utilize larger amounts of calcium and lesser amounts of potassium than non-legumes. Is it not possible that the presence of symbiotic bacteria in the roots of legumes may have some influence on the absorption and utilization of these and other nutrient ions? It was in the hope of obtaining additional information on these points that this study was undertaken.

Plan and Methods

A better understanding of the properties and behavior of colloidal clay has made possible the use of this fraction of the soil in the preparation of cultural media (23, 28). Furthermore, the success from the use of such media has been demonstrated (5, 7, 14, 17, 21).

Preparation of Colloidal Clay—The colloidal clay used in this study was prepared in the manner described by Bradfield (8). A 5 per cent suspension of the electrodialyzed clay had a pH of 3.7 as determined with a glass electrode potentiometer. The acidic properties of the clay are shown by the curve of Fig. 1. The colloidal clay possessed an exchange capacity of 66 milligram equivalents (M. E.) per one hundred grams of the clay.

Preparation of Cultures—The method of preparing the cultural media has been given, among others, by Horner (21), and Ferguson (14). Two sets of soybeans were each supplied three constant amounts of potassium and each potassium level combined with two different levels of calcium. The nutrient ions, as given in Table 1, were titrated on to the colloidal clay so as to supply them in exchangeable forms. Sufficient barium was added as a supplementary ion to each culture to maintain a constant degree of base saturation of the clay. This clay was then mixed with sufficient

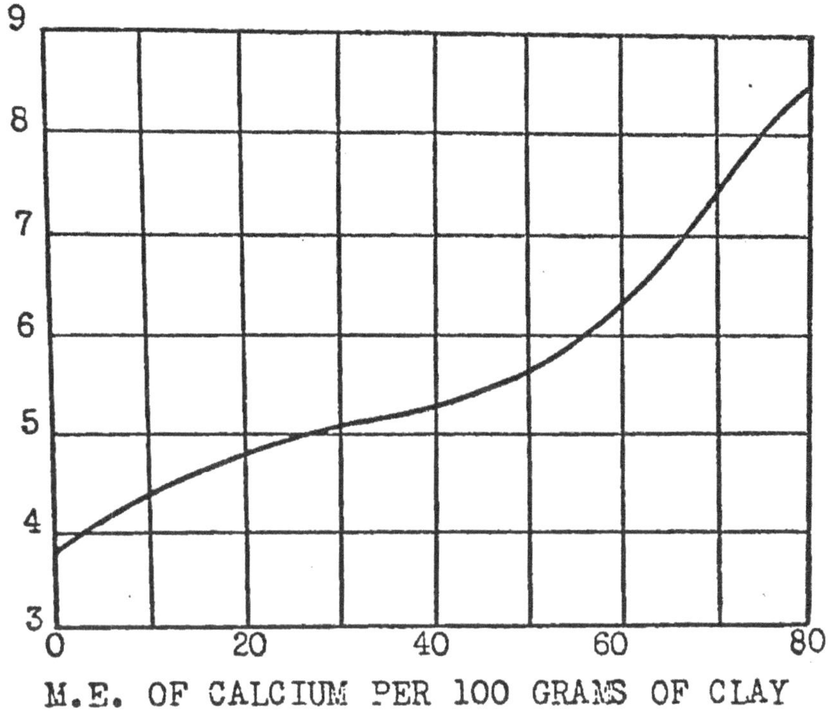

Fig. 1. Curve of hydrogen clay by calcium hydroxide.

acid-washed quartz sand to insure desirable physical conditions for plant growth.

Two groups of soybeans, designated as Series A and Series B, were grown. Series B consisted of the nodulated soybeans and Series A was grown without bacterial infection. Soybeans of the Virginia variety were used. Each treatment was grown in duplicate for a period of five weeks. The two crops were grown in the greenhouse during May and early June of succeeding years. At the end of the growth period the tops and roots of the plants were harvested separately, dried, weighed, and then were finely ground and mixed together.

TABLE 1: Ions added to clay cultures by titration.					
Series A (Used for three successive crops)					
Culture No.	Ions Added (Milligram Equivalents)				
	K	Ca	Mg	P	Ba
1	5	10	10	7.5	25
2	10	10	10	7.5	20
3	15	10	10	7.5	15
4	5	20	10	7.5	15
5	10	20	10	7.5	10
6	15	20	10	7.5	5
Series B (Used for a single crop)					
1	5	10	10	7.5	20
2	10	10	10	7.5	15
3	15	10	10	7.5	10
4	5	20	10	7.5	10
5	10	20	10	7.5	5
6	15	20	10	7.5	0

Analytical Methods—Samples of the plant material were digested by heating with a mixture of nitric and perchloric acids (16). The barium, calcium, magnesium, potassium, and phosphorus were determined in aliquots of the hydrochloric acid solution of the ash using accepted methods (18, 24, 31).[2] The ash residue was purified (31) and reported as silicon. The nitrogen determinations were made by the modified Kjeldahl method (29). The analysis of the fractions—total sugars, and starch and hemicellulose combined— were made according to a modification of the method of Shaffer and Somogyi (19). The residues from the carbohydrate determination were used for the estimation of cellulose and lignin by the procedure described by Crampton (11). The analytical data represent the analysis in duplicate of the combined tops and roots of fifty plants.

[2] *Credit is due Dr. V. R. Ells for the spectrographic determination of magnesium and potassium in a portion of the plant material.*

Experimental Results: Legume Behavior Crop One, Series B

Growth of Soybeans—The rate of maturation of the soybean plants was progressively greater as increasing amounts of potassium were supplied. At the time of harvest, the cotyledons of the plants to which 15 M. E. of potassium were supplied had either fallen or were completely yellow. In cultures supplied with 5 M. E. of adsorbed potassium, the cotyledons showed little evidence of translocation. The plants supplied with 15 M. E. of potassium had slender, weak stems. The calcium level seemed to exert little influence on the strength of the stems. The leaves of the plants growing in cultures treated with 5 M. E. of potassium showed distinct symptoms of potassium deficiency, but the plants of cultures supplied with 15 M. E. of potassium showed only slight symptoms of potassium shortage. The plants of the cultures treated with 10 M. E. of potassium were intermediate in appearance between those of the low and the high potassium cultures.

The growth weights obtained in the plants of this group were approximately proportional to the amounts of potassium supplied, whether at the low or the high calcium level. The cultures supplied with 5, 10, and 15 M. E. of potassium and 10 M. E. of calcium produced 9.907, 11.402, and 14.267 grams of oven-dried plant material, according to the data for weights in Table 2. The cultures to which the same increments of potassium were added but combined with 20 M. E. of calcium produced 13.200, 13.945, and 15.255 grams of plant material respectively.

TABLE 2: Growth of soybeans in relation to variable amounts of potassium and calcium. (Data for 50 plants. Crop 1, Series B)						
Culture No.	Treatments		Weight of 50 plants			Ratio Tops Roots
	M.E. Potassium	M.E. Calcium	Tops (Gms.)	Roots (Gms.)	Total (Gms.)	
1	5	10	7.655	2.302	9.907	3.32
2	10	10	9.140	2.262	11.402	4.04
3	15	10	11.170	3.097	14.267	3.61
4	5	20	10.235	2.965	13.200	3.45
5	10	20	10.785	3.160	13.945	3.41
6	15	20	11.923	3.332	15.255	3.58

Increasing the supply of potassium at the lower calcium level brought about a more rapid increase in plant weight than the same increase in the potassium supply at the higher calcium level, as shown in Fig. 2. These results point to the influence of potassium in stimulating vegetative growth when the calcium supply is limited.

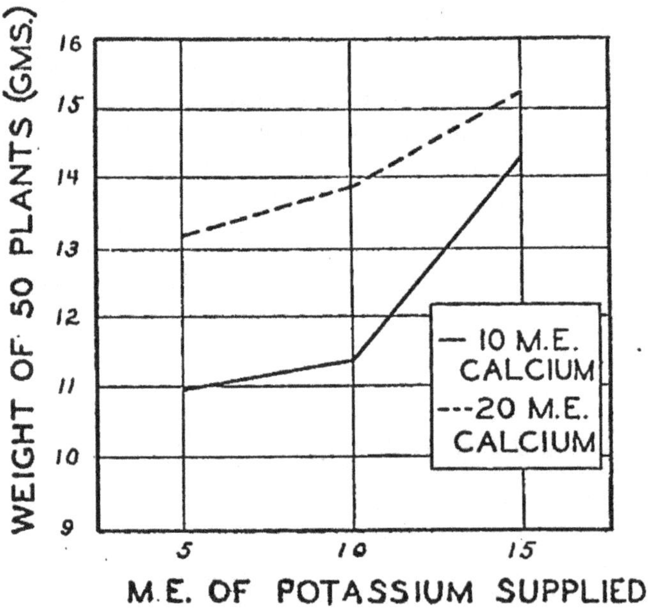

Fig. 2. Growth of soybeans as influenced by variable amounts of potassium and calcium. (Crop 1, Series B).

At each amount of potassium, a higher yield of plant material was obtained at the higher calcium level, as indicated by the graphic presentation of the data in Fig. 2. The difference between the weights of oven-dried plant material produced at the two calcium levels was greatest when the amount of potassium supplied was 5 M. E., and least when it was 15 M. E. Thus calcium exerted a greater influence on growth when the available supply of potassium was low. Not only did greater total plant growth accompany the larger calcium supply, but there was a corresponding increase in root weight as well as top weight.

Nitrogen Fixation—Both the potassium and calcium influenced the nitrogen fixation of soybeans grown in colloidal clay-sand cultures and supplied with variable levels of these nutrient cations. The concentration of nitrogen in the plant material declined as the potassium supply increased. The cultures supplied with 5, 10, and 15 M. E. of potassium and 10 M. E. of calcium contained 3.99, 3.30, and 3.24 per cent nitrogen respectively. Those treated with the same amounts of potassium but with 20 M. E. of calcium contained 4.00, 3.43, and 3.04 per cent nitrogen. The data are given in Table 3. The decrease in the concentration of nitrogen in the plant tissue was affected only slightly by the calcium supplied. The trend was the same at the low and at the high calcium levels. The nitrogen in the plant material expressed as percentage of dry weight was a direct reciprocal of vegetative growth. Since a greater potassium supply brought about increased growth, the amount of nitrogen per unit weight of plant tissue was reduced although the total nitrogen in fifty plants increased with the potassium supply.

TABLE 3: Nitrogen fixation in soybeans in relation to variable amounts of potassium and calcium.

(Data for 50 plants. Crop 1, Series B)

Culture No.	Treatments		Number of Nodules per 50 plants	Nitrogen*		
	M.E. Potassium	M.E. Calcium		Per Cent	Total Mgms.	Fixed Mgms.
1	5	10	151	3.99	394.7	117.7
2	10	10	217	3.30	376.4	99.4
3	15	10	347	3.24	462.4	185.4

4	5	20	273	4.00	519.9	242.9
5	10	20	318	3.43	479.7	202.7
6	15	20	438	3.04	463.4	186.4
*Fifty seeds had 277 mgms. nitrogen.						

The cultures treated with 20 M. E. of calcium produced a greater yield and a greater total amount of nitrogen in the crop, but the concentration of nitrogen in the plants declined with increasing potassium with the low value of this series where a high calcium supply was associated with the highest level of potassium. This was not quite as low as the highest in the series with 10 M. E. of calcium.

Nitrogen fixation at the low calcium level increased from the lowest to the highest level of potassium, though as shown in Table 3, the amount of nitrogen in the culture supplied with 10 M. E. of potassium was less than that in the other two. When 20 M. E. of calcium were supplied, the nitrogen fixation and the total nitrogen declined as the potassium supply increased. Cultures treated with 5, 10, and 15 M. E. of potassium and 10 M. E. of calcium contained respectively 394.7, 376.4, and 462.4 mgms. of nitrogen in fifty plants. The plants to which 20 M. E. of calcium were available at the same potassium levels contained 519.9, 479.7, and 463.4 mgms. of total nitrogen.

The marked beneficial effect of a greater calcium supply on nitrogen fixation is shown by the graphic presentation of the data in Fig. 3. This influence was particularly significant at low potassium levels where the ratio of the calcium to the potassium supplied was wide. At the high potassium level there was little difference between the amount of nitrogen fixed in cultures supplied 10 M. E. and 20 M. E. of adsorbed calcium. The fact that the number of nodules on the cultures increased with additions either of calcium or potassium indicates the favorable effect of each of these nutrient cations on bacterial infection, although the calcium supply influenced the efficiency of the microorganisms in nitrogen fixation the more significantly.

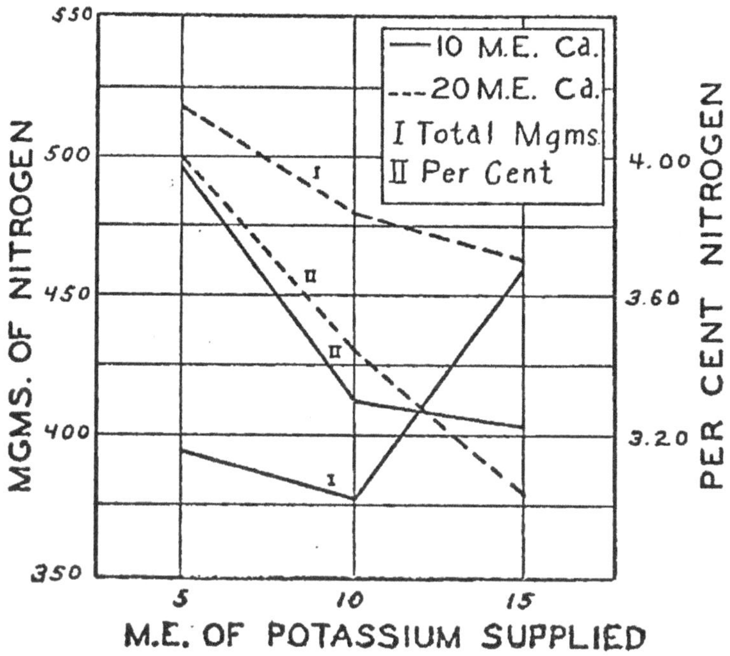

Fig. 3. Nitrogen content of soybeans as influenced by variable amounts of potassium and calcium. (Crop 1, Series B).

Carbohydrate Contents—The concentration of total sugars in the plant materials decreased as the supply of potassium increased. This is shown in the data assembled in Table 4. In the cultures treated with 5, 10, and 15 M. E. of potassium and 10 M. E. of calcium, the plant tissue contained 1.80, 1.60, and 1.34 per cent total sugars respectively. The concentrations in the cultures supplied with the same levels of potassium along with 20 M. E. of calcium were 1.42, 1.41, and 0.90 per cent. Again because of increased growth, the total amount of sugars found in fifty plants increased as the supply of potassium increased, with the exception of the culture treated with 15 M. E. of potassium and 20 M. E. of calcium.

A greater calcium supply resulted in a slight decrease in the concentration of total sugars; however, there was no consistent difference in the total amounts of sugars in the plants at the two calcium levels.

Although there was little difference due to treatment in the fraction composed of starch and hemicellulose, the total lignin and cellulose in the plant materials increased with increments of potassium at both calcium levels.

In general, the ratio of the available carbohydrate carbon to nitrogen in the plant tissue increased as the supply of potassium increased, as shown in Table 4. It is particularly interesting that the narrowest ratio of carbon to nitrogen was found in the culture in which the narrowest ratio of potassium to calcium was supplied. That the increments of potassium increase the nitrogen fixation has been shown in Table 3, but the data of Table 4 indicate that increased amounts of exchangeable potassium influence more markedly the production of carbohydrates in the plants. The level of calcium supplied to the plants had no significant influence on the ratio of the carbon to the nitrogen in the plant material. The relationship is shown graphically in Fig. 4.

TABLE 4: Carbohydrate contents of plant material in relation to variable amounts of potassium and calcium.

(Data for 50 plants. Crop 1, Series B)

Culture No.	Total sugars		Starch + Hemicellulose		Cellulose		Lignin		Ratio C*/N	Ratio K/Ca Supplied
	Per Cent	Total Mgms.	Per Cent	Total Mgms.	Per Cent	Total Mgms.	Per Cent	Total Mgms.		
1	1.80	178	17.23	1714	20.57	2039	8.95	887	5.8	1
2	1.60	183	13.57	1550	22.89	2607	10.24	1166	7.8	2
3	1.34	191	13.08	1869	21.40	3052	9.43	1345	6.8	3
4	1.42	189	15.30	2027	24.14	3187	8.22	1123	5.6	0.5
5	1.41	195	12.27	1713	23.43	3068	10.16	1439	6.4	1
6	0.90	137	15.01	2286	22.04	3362	10.46	1610	7.8	1.5

* Carbohydrate carbon only was used in calculating this ratio.

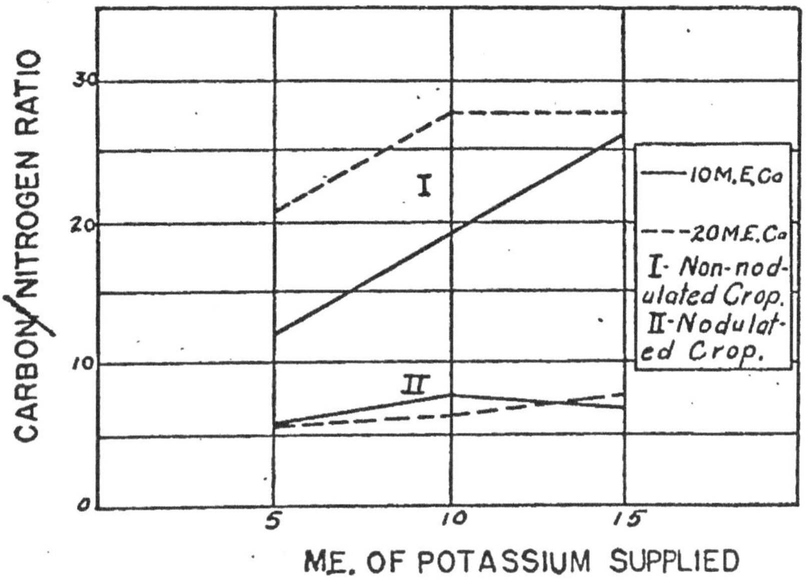

Fig. 4. Ratios of carbohydrate carbon to nitrogen in soybeans as influenced by variable amounts of potassium and calcium and by nitrogen fixation.

Potassium Contents—The concentration of potassium in the plant materials showed a direct relation to the amount of potassium supplied. The cultures treated with 5, 10, and 15 M. E. of potassium and 10 M. E. of calcium contained respectively 2.85, 3.87, and 4.34 percent of potassium, as given in Table 5. Cultures offered the same amounts of adsorbed potassium but supplied with 20 M. E. of calcium contained 2.13, 3.30, and 3.96 percent of potassium. The lower concentration of potassium at the higher calcium level may be explained chiefly by the increased growth induced by the additional calcium.

TABLE 5: Composition of soybeans as influenced by variable amounts of potassium and calcium.										
(Data for 50 plants. Crop 1, Series B)										
Culture No.	Treatments		Potassium		Calcium		Magnesium		Ratio K/Ca Absorbed by the Plants
	M.E. Potassium	M.E. Calcium	Per Cent	Total Mgms.	Per Cent	Total Mgms.	Per Cent	Total Mgms.	
1	5	10	2.85	282.4	0.40	39.1	0.588	58.2	7.2
2	10	10	3.87	442.1	0.53	47.0	0.455	51.9	9.4
3	15	10	4.34	620.0	0.38	54.6	0.356	50.9	11.3
4	5	20	2.13	280.9	0.70	91.9	0.668	88.3	3.1
5	10	20	3.30	460.2	0.70	98.3	0.463	64.8	4.7
6	15	20	3.96	600.6	0.67	101.5	0.379	57.8	5.9

The total quantity of potassium in the plant tissue was primarily a matter of the potassium supply offered by the soil. The cultures supplied with 5, 10, and 15 M. E. of potassium and 10 M. E. of calcium contained respectively 282.4, 442.1, and 620.0 mgms. of potassium. The plants of cultures given the same potassium treatments but supplied with 20 M. E. of calcium contained 280.9, 460.2, and 600.6 mgms. of potassium. That the potassium content is almost a linear function of the supply is evident from the graphic presentation of the data in Fig. 5. The potassium content of the plants was very slightly lower at the higher calcium level.

The portion of the available potassium taken by the plants of each culture increased somewhat as higher levels of potassium were offered, as shown by the data of Table 6. In all cultures, however, the plants took up a rather large percentage of the potassium available. Cultures treated with 5, 10, and 15 M. E. of potassium and 10 M. E. of calcium utilized 76.1, 78.2, and 81.5 per cent of the potassium supplied. The plants consumed 75.8, 81.4, and 79.0 per cent of the available potassium in cultures supplied with the same amounts of potassium but with 20 M. E. of calcium. Although the plants which were offered 5 M. E. of potassium utilized this nutrient element nearly as efficiently as the plants supplied with higher levels, the appearance of the plants indicated pronounced symptoms of potassium deficiency.

TABLE 6: Consumption of potassium, calcium, and magnesium in relation to the total amounts supplied.									
(Data for 50 plants. Crop 1, Series B)									
Culture No.	Potassium			Calcium			Magnesium		
	Supplied Mgms.	Taken		Supplied Mgms.	Taken		Supplied Mgms.	Taken	
		Mgms.	Per Cent		Mgms.	Per Cent		Mgms.	Per Cent
1	371.5	282.4	76.1	209.9	39.1	18.6	132.1	58.2	44.1
2	566	442.1	78.2	209.9	47.0	22.4	132.1	51.9	39.3
3	761.5	620.0	81.5	209.9	54.6	26.1	132.1	50.9	38.5
4	371.5	280.9	75.8	409.9	91.9	22.4	132.1	88.3	66.8
5	566	460.2	81.4	409.9	98.3	24.0	132.1	64.8	49.1
6	761.5	600.6	79.0	409.9	101.5	24.8	132.1	57.8	43.8

*Amounts supplied include that by titration and in seeds. Fifty seeds contained 176 mgms. Potassium, 9.9 mgms. Calcium, and 8.9 mgms. Magnesium.

There was little difference between the efficiency of potassium consumption because of the difference in the calcium levels, according to the graph in Fig. 5.

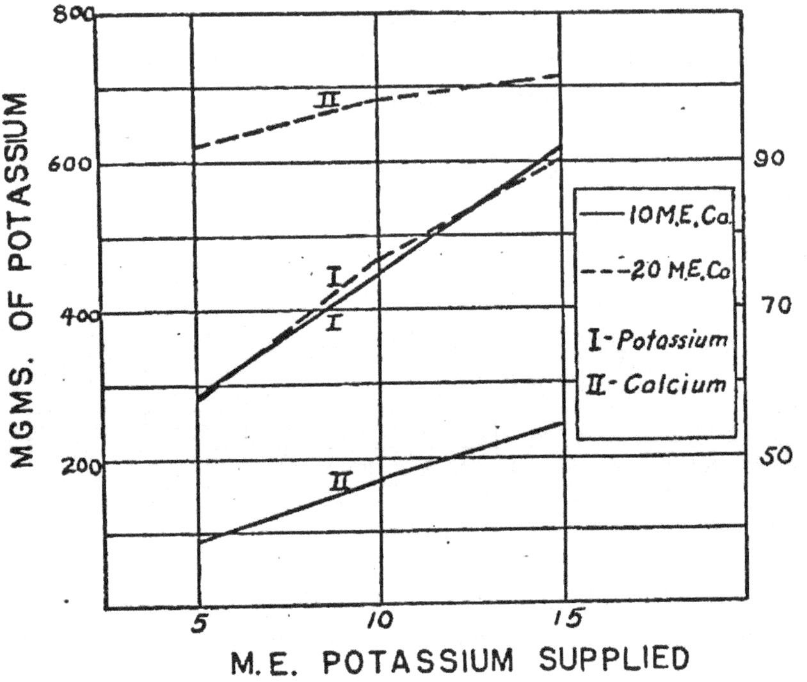

Fig. 5. Potassium and calcium contents of soybean plants as influenced by variable amounts of potassium and calcium. (Crop 1, Series B).

Calcium Contents—The concentrations of calcium in the plant tissue, according to the data of Table 5, were roughly constant at each level of calcium offered. In the cultures supplied with 20 M. E. of calcium, however, the concentrations were considerably higher than in those given 10 M. E. of calcium. The concentrations declined somewhat at higher levels of potassium even when the calcium supply was doubled.

The total amounts of calcium taken by the plants, at both calcium levels, increased as larger quantities of potassium were available, as shown by the data of Table 5. The cultures supplied with 5, 10, and 15 M. E. of potassium and 10 M. E. of calcium contained 39.1, 47.0, and 54.6 mgms. of calcium respectively. Cultures given the same potassium treatments but supplied with 20 M. E. of calcium contained respectively 91.9, 98.3, and 101.5 mgms. of calcium. The total calcium contained in fifty plants was increased by additions of both potassium and calcium, as shown by the graph in Fig. 5. However, a higher calcium supply resulted in a much greater increase in the calcium content.

The total calcium in the plant tissue represents 18.6, 22.4, and 26.1 per cent of thc constant supply of 209.9 mgms. of calcium offered in cultures 1, 2, and 3; and 22.4, 24.0, and 24.8 per cent of the 409.9 mgms. offered in cultures 4, 5, and 6, as shown in Table 6. It is interesting that the increased efficiency of calcium utilization by the plants was associated with increased root weight occasioned by higher nutrient levels as increments of potassium were available.

Phosphorus Contents—In general, the concentrations of phosphorus in the plant materials declined as increments of potassium were offered. This is shown in Table 7. The cultures treated with 5, 10, and 15 M. E. of potassium and 10 M. E. of calcium contained respectively 0.454, 0.418, and 0.378 per cent of phosphorus. With the same levels of potassium but coupled with 20 M. E. of calcium, the plants contained 0.383, 0.323, and 0.330 per cent of phosphorus. An increased calcium supply resulted in a decrease in the concentration of phosphorus. The decline in phosphorus concentration, which was occasioned by additions of both calcium and potassium, can be attributed to the increased yields of plant material as a result of higher fertility levels. Although increments of potassium supplied to the plants brought about no significant difference in the concentration of calcium at each calcium level, increasing supplies of potassium were associated with decreasing concentrations of phosphorus.

TABLE 7: Plant contents of phosphorus, silicon, and barium in relation to variable potassium and calcium.

(Data for 50 plants. Crop 1, Series B)

Culture No.	Treatments		Phosphorus		Silicon		Barium	
	Potassium M. E.	Calcium M. E.	Per Cent	Total Mgms.	Per Cent	Total Mgms.	Per Cent	Total Mgms.
1	5	10	0.454	44.4	2.54	251.2	1.52	151.4
2	10	10	0.418	47.8	2.15	246.0	0.65	74.4
3	15	10	0.378	53.9	1.46	208.2	0.28	39.4
4	5	20	0.383	50.5	1.60	211.0	0.36	47.6
5	10	20	0.323	45.9	1.08	151.1	0.22	30.2
6	15	20	0.330	50.4	0.81	124.5	0.00	0.0

Increments of potassium increased, however, the total phosphorus contents of the plants in the cultures treated with 10 M. E. of calcium. Cultures which offered the same supplies of potassium with 20 M. E. of calcium grew plants which were roughly equal in contents of total phosphorus, and not so widely different from those given 10 M. E. of calcium. Plants supplied with 5, 10, and 15 M. E. of potassium and 10 M. E. of calcium contained respectively 44.4, 47.8, and 53.9 mgms. of phosphorus. Those offered the same amounts of potassium supplied at the higher calcium level contained 50.5, 45.9, and 50.4 mgms. of phosphorus.

That the utilization of the phosphorus offered to the plants in colloidal clay cultures is more efficient as the ratio of the potassium to the calcium in the medium becomes wider is indicated by the data obtained in this work to substantiate those reported by Ferguson (14). In the cultures supplied with 10 M. E. of calcium, the plants used 39.0, 42.0, and 47.4 per cent of the phosphorus offered respectively at levels of 5, 10, and 15 M. E. of potassium. The ratios of potassium to calcium offered were in the order 1, 2, 3. The plants of cultures treated with 20 M. E. of calcium and the same potassium levels presented narrower ratios of potassium to calcium and the per cent of the phosphorus supplied which was utilized by the plants was more nearly constant.

The greater intake of phosphorus from a constant supply with additions of both calcium and potassium again suggests that increased root area is a factor in nutrient absorption by plants.

Magnesium Contents—The concentrations of magnesium in the plant materials decreased regularly with additions of potassium, as shown by the data of Table 5. The plant tissue from cultures supplied with 5, 10, and 15 M. E. of potassium and 10 M. E. of calcium contained 0.588, 0.455, and 0.356 per cent of magnesium respectively; and that from cultures treated with the same levels of potassium but with 20 M. E. of calcium contained 0.668, 0.463, and 0.379 per cent of magnesium. While more exchangeable potassium decreased magnesium concentration, a higher calcium supply resulted in a greater concentration of magnesium. Magnesium in total amounts decreased regularly with increments of potassium, as is indicated by the data of Table 5. Plant tissue from cultures which offered 5, 10, and 15 M. E. of potassium and 10 M. E. of calcium contained respectively 58.2, 51.9, and 50.9 mgms. of magnesium. That from treatments with the same levels of potassium combined with 20 M. E. of calcium had 88.3, 64.8, and 57.8 mgms. of magnesium.

The magnesium content of the plant material, expressed either as percentage or as total amount in fifty plants, was increased by a greater calcium supply. The data are presented graphically in Fig. 6. The absorption by the plant of the magnesium supplied by the seed and by titration decreased as the amounts of potassium offered increased, but the utilization by the plants of the available magnesium was more efficient at the higher calcium level.

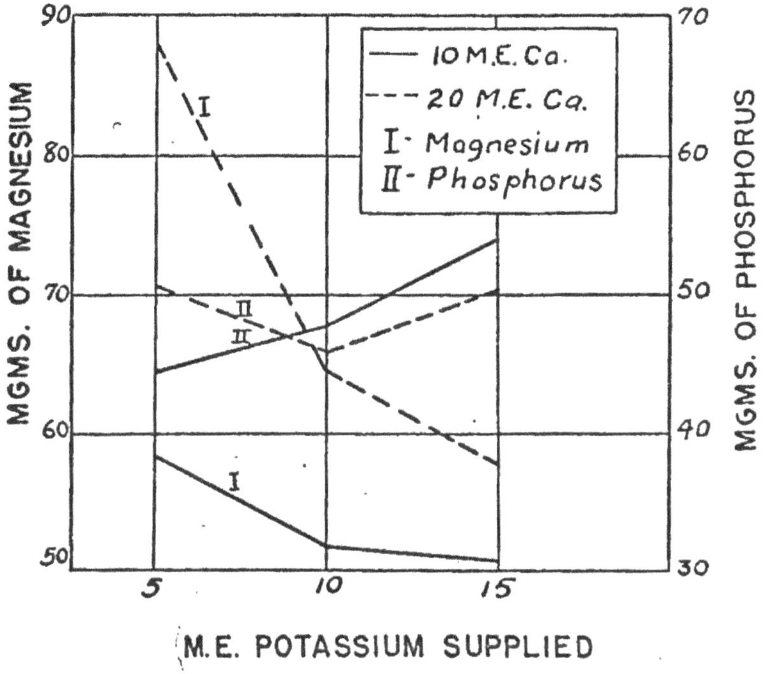

Fig. 6. Magnesium and phosphorus contents of soybeans as influenced by variable amounts of potassium and calcium.

Barium and Silicon Contents—Inasmuch as barium and silicon are not plant nutrients, they are classed together. Even though barium is a cation, whereas silicon functions as an anion, their behaviors as influenced by varying potassium and calcium treatments were similar. The amounts of each as concentrations and as totals contained in the plant tissue declined as increments of potassium were supplied.

Barium was supplied in amounts reciprocal to those of the variable cations—potassium and calcium. The amount of barium adsorbed on the colloidal complex, therefore, decreased with increments of the potassium and calcium. The barium contents of the plant tissue, expressed either as percentage or as total milligrams per culture, were directly proportional to the amounts supplied. Cultures treated with 5, 10, and 15 M. E. of potassium and 10 M. E. of calcium contained 1.52, 0.65, and 0.28 per cent of barium respectively. In cultures treated with the same levels of potassium but with 20 M. E. of calcium, the concentrations of barium were found to be 0.36, 0.22, and 0.00 per cent. Those data are given in Table 7, and the barium treatments are shown in Table 8.

TABLE 8: Consumption of barium, phosphorus, and nitrogen in relation to the total amounts supplied.*

(Data for 50 Plants. Crop 1, Series B)

Culture No.	Barium			Phosphorus			Nitrogen		
	Supplied Mgms.	Taken		Supplied Mgms.	Taken		Supplied Mgms.	Taken	
		Mgms.	Per Cent		Mgms.	Per Cent		Mgms.	Per Cent**
1	1373.6	151.4	11.0	113.7	44.4	39.0	277	394.7	142.5
2	1030.2	74.4	7.2	113.7	47.8	42.0	277	376.4	135.9
3	686.8	39.4	5.8	113.7	53.9	47.4	277	462.4	166.9
4	686.8	47.6	6.9	113.7	50.5	44.4	277	519.9	187.6
5	343.4	30.2	8.8	113.7	45.9	40.4	277	479.7	173.2
6	0.0	0.0	-	113.7	50.4	44.4	277	463.4	167.3

*Fifty seeds contained 0.00 mgms. Barium, 36.2 mgms. Phosphorus, and 277 mgms. Nitrogen.
**In the case of nitrogen there was an increase over the amount supplied because of nitrogen fixation.

The total barium contained in the plant tissue was likewise found to be directly proportional to the amount adsorbed on the clay. Cultures treated with 5, 10, and 15 M. E. of potassium and 10 M. E, of calcium contained respectively 151.4, 74.4, and 39.4 mgms. of barium. The cultures supplied with the same amounts of potassium but with 20 M. E. of calcium contained 47.6, 30.2, and 0.0 mgms. of barium respectively.

The silicon content of the plant tissue was very similar to that of the barium. As the amounts of nutrient cations adsorbed on the colloidal complex increased the silicon content of the plants decreased. As increments of potassium were supplied at the low calcium level, the concentration of silicon in the plant tissue decreased from 2.54 to 1.46 per cent; and at the high calcium level, silicon decreased from 1.60 to 0.81 per cent, as shown by the data of Table 7. Cultures treated with 5, 10, and 15 M. E. of potassium and 10 M. E. of calcium contained a total of 251.2, 246.0, and 208.2 mgms. of silicon. The plant tissue from cultures supplied with the same potassium levels but with 20 M. E. of calcium contained respectively 211.0, 151.1, and 124.5 mgms. of silicon.

A higher calcium supply resulted in a significant reduction in the silicon content, expressed either as percentage or as total amount, as the graphic presentation of the data in Fig. 7, shows.

It is interesting to note, as the data show, that the silicon delivery to the plants was most pronounced in the cultures supplied with the lowest amounts of nutrient ions. The absorption of barium, however, appeared to be governed more by the amount of barium adsorbed on the clay than by the direct influence of the nutrient cation level of the medium, but again more barium was on the clay as there was less of the nutrient ions.

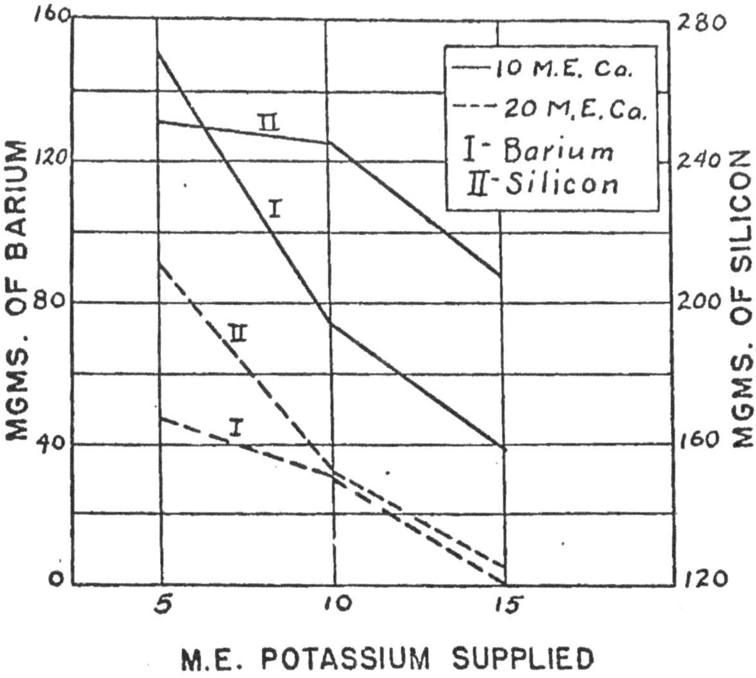

Fig. 7. Barium and silicon contents of soybeans as influenced by variable amounts of potassium and calcium. (Crop 1, Series B).

Non-Legume Behavior Crop One, Series A

One crop of the soybeans was not nodulated and although a legume taxonomically, it functioned physiologically as a non-legume. In plant performance and behavior this crop served as a means of comparing the other crop as legume with this one of non-leguminous plants.

Growth of Non-nodulated Soybeans—Although the plants of this series were supplied only the nitrogen contained in the seeds, good growth weights were obtained. The weights of oven-dry plant material increased as increments of potassium were supplied. In the cultures treated with 5, 10, and 15 M. E. of potassium and 10 M. E. of calcium, those weights were 12.865, 19.790, and 19.885 grams respectively. With the same potassium supplies but growing at the higher calcium levels, the dried plant materials weighed 18.955, 22.045, and 22.420 grams. These data are given in Table 9.

Additions of calcium, likewise, brought about increased production of plant material, as shown by the graph of Fig. 8. As in Series B, with leguminous behaviors, so in this Series A with non-leguminous behaviors, the calcium exhibited the greatest influence on growth when associated with the lowest potassium level. Moreover, increasing the supply of potassium at the lower calcium level resulted in a more rapid increase in the plant weight than the same increments of potassium coupled with the higher calcium level. There was little difference between the ratios of the tops to the roots in the several cultures of this series.

TABLE 9: Growth of soybeans in relation to variable amounts of potassium and calcium.

(Data for 50 plants. Crop 1, Series A)

Culture No.	Treatments		Weight of 50 Plants			Ratio Tops Roots
	Potassium M. E.	Calcium M. E.	Tops (Gms.)	Roots (Gms.)	Total (Gms.)	
1	5	10	8.035	4.820	12.855	1.67
2	10	10	14.075	5.715	19.790	2.43
3	15	10	13.815	6.070	19.885	2.28
4	5	20	13.060	5.895	18.955	2.22
5	10	20	16.015	6.030	22.045	2.66
6	15	15*	16.270	6.150	22.420	2.64

Fifty seeds contained 346 mgms. of nitrogen
* Because the plant material of the culture of Crop 1, Series A initially treated with 15 M.E. of potassium and 20 M.E. of calcium was lost, the plant material from a culture treated with 15 M.E. of potassium and 15 M.E. of calcium was substituted.

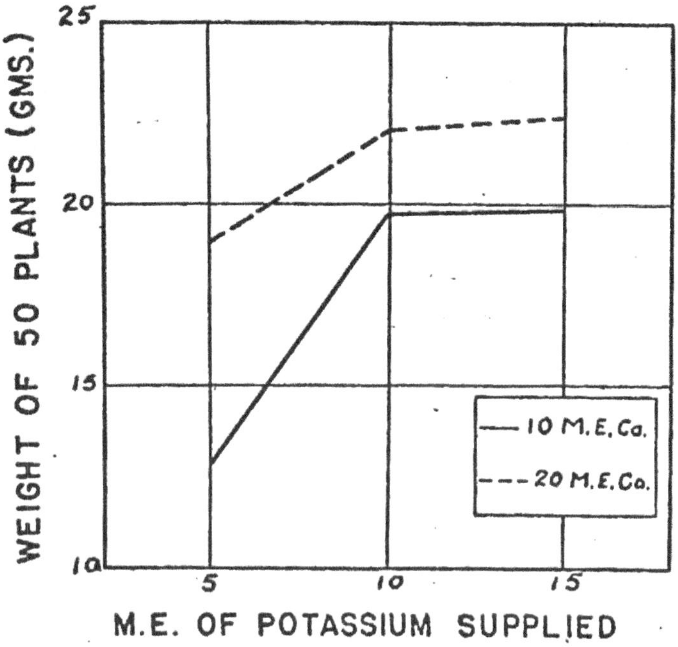

Fig. 8. Growth of soybeans as influenced by variable amounts of potassium and calcium. (Crop 1, Series A).

At the time of harvest, the maturation of the plants was progressively proportional to the potassium supplied. The cotyledons of the plants treated with 5 M. E. of potassium showed only moderate translocation and the lower leaves were green. All of the plants were erect. On the other hand, both the cotyledons and the lower leaves of plants grown in cultures supplied with 15 M. E. of potassium were quite yellow, showing pronounced translocation of certain constituents, particularly nitrogen, to the upper portions of the plants. The plants of these cultures were slightly lodged, an indication of weak structural material in the stems. Plate 1 gives a photographic presentation of the appearance. The plants growing in cultures treated with 10 M. E. of potassium were intermediate in appearance.

Plate 1. Growth of soybeans influenced by potassium and calcium. (Crop one, Series A. Increasing potassium from left to right. A represents 10 M. E. and B represents 20 M. E. of calcium.)

The difference in the levels of calcium supplied the plants had little effect on translocation and strength of stems. The leaves of the plants growing in cultures supplied with 5 M. E. of potassium showed distinct symptoms of potassium deficiency.

Nitrogen Content of Non-nodulated Soybeans—This crop was not inoculated with nitrogen fixing bacteria and nitrogen was not added to the cultures. The nitrogen content of the plant material, expressed as a percentage of dry weight, decreased directly with increased growth. The total nitrogen content, however, was approximately constant, and closely approximated the nitrogen content of the seeds. The pronounced translocation which occurred in the plants supplied with the higher levels of potassium indicated that nitrogen was the factor limiting the growth. That the plants of this series utilized a higher percentage of the available nitrogen is evident from the figures given in Table 15. Other nitrogen features appear in Table 10.

TABLE 10: Nitrogen contents of soybeans in relation to variable amounts of potassium and calcium.

(Data for 50 plants. Crop 1, Series A)

Culture No.	Treatments		Nitrogen		Ratio C : N	Ratio K : Ca Supplied
	Potassium M. E.	Calcium M. E.	Per Cent	Total Mgms.		
1	5	10	2.66	341.2	8.5	0.5
2	10	10	-	-	-	1.0
3	15	10	1.70	347.8	18.6	1.5
4	5	20	2.21	384.5	13.9	0.25
5	10	20	1.51	341.2	18.0	0.5
6	15	15*	1.50	334.2	18.5	1.0

Fifty seeds contained 346 mgms. of nitrogen.

*Because the plant material of the culture of Crop 1, Series A initially treated with 15 M.E. of potassium and 20 M.E. of calcium was lost, the plant material from a culture treated with 15 M.E. of potassium and 15 M.E. of calcium was substituted.

Carbohydrate Contents—The various carbohydrate fractions of the plant tissue followed closely the increased growth induced by additions of potassium. The rates of carbohydrate synthesis and storage, according to the data of Table 11, were greatly increased by increments of potassium when the nitrogen supply was low. That the total carbohydrate content was considerably higher in the plant material of the non-nodulated crop is evident from a comparison of the data assembled in Table 4 with that of Table 11.

A particularly significant difference occurred in the case of total sugars. As increasing supplies of potassium were offered with 10 M. E. of calcium, the amount of total sugars increased from 529 to 904 mgms. In cultures treated with the same amounts of adsorbed potassium but with the higher calcium level, the amount of total sugars contained in fifty plants increased from 524 to 1073 mgms. In comparison, the content of total sugars in the nodulated crop of Series B varied from a low value of 137 to a high of 195 mgms. This suggests that the greater nitrogen content of the nodulated soybeans brought about a rapid utilization of the simple sugars in protein synthesis. That a greater storage of carbohydrates occurred in the non-nodulated crop during plant growth as a result of the lack of nitrogen is likewise suggested by the data.

TABLE 11: Carbohydrate content of plant material in relation to variable amounts of potassium and calcium.

(Data for 50 plants. Crop 1, Series A)

Culture No.	Total Sugars		Starch + Hemicellulose		Cellulose		Lignin	
	Per Cent	Total Mgms.	Per Cent	Total Mgms.	Per Cent	Total Mgms.	Per Cent	Total Mgms.
1	4.12	529	9.90	1270	18.71	2399	11.52	1478
2	-	-	-	-	-	-	-	-
3	4.42	904	18.73	3834	23.19	4744	17.10	3499
4	3.01	524	8.97	1842	27.54	4788	19.10	3325
5	4.83	1073	17.08	3796	25.28	5620	11.71	2604
6	3.81	851	15.62	3508	27.74	5771	10.4	2331

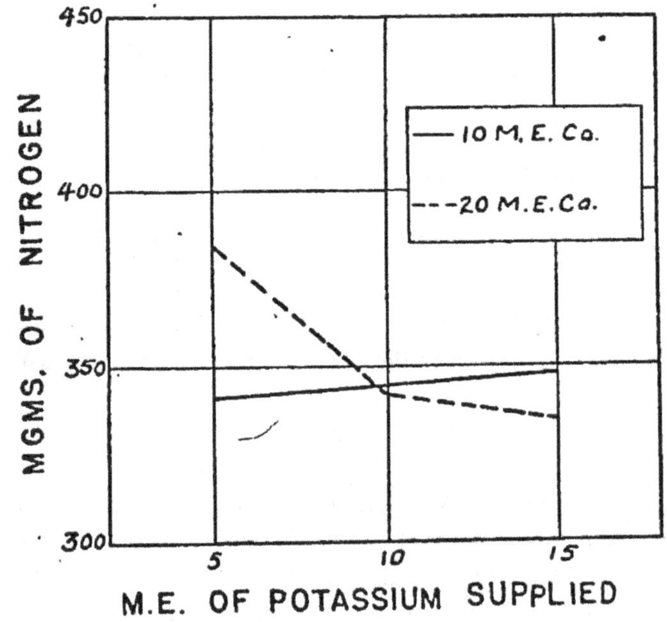

Fig. 9. Nitrogen content of soybeans as influenced by variable amounts of potassium and calcium. (Crop 1, Series A).

The ratio of the carbohydrate carbon to the nitrogen in the plant tissue widened as increased amounts of potassium were available to the plants, as is shown by the calculations assembled in Table 10. At the low calcium level, the ratio increased from 8.5 in the plants supplied with 5 M. E. of potassium to 18.6 in those offered 15 M. E. When a greater amount of calcium was supplied, the ratios increased from 13.9 at the low to 18.5 at the high potassium level.

That there is a significant difference between the ratios of carbohydrate carbon to nitrogen of the plant tissues of nodulated and non-nodulated soybeans is shown by a comparison of the values given in Table 16 and in Fig. 10.

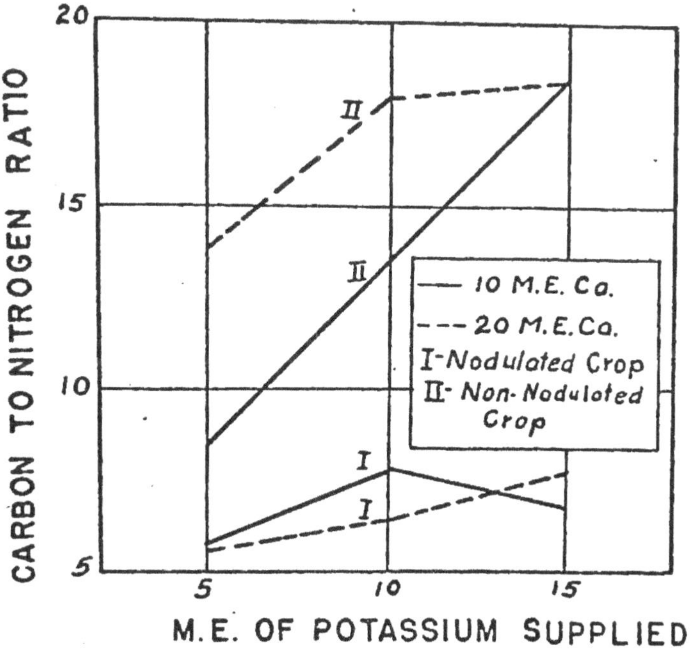

Fig. 10. Ratios of carbohydrate carbon to nitrogen in soybeans as influenced by nitrogen fixation and variable amounts of potassium and calcium.

Potassium Contents—The concentrations of potassium in the plant tissues of the non-nodulated soybeans increased as larger amounts of potassium were available. The cultures treated with 5, 10, and 15 M. E. of potassium and 10 M. E. of calcium contained 1.83, 2.07 and 2.41 per cent respectively of potassium. Those to which the same potassium treatments were supplied at the higher calcium level contained 1.33, 1.80, and 2.10 per cent of potassium. The data are shown in Table 12. At all potassium levels, the concentration of potassium was lowered by a higher calcium supply, but the lower percentage may be explained by the increased growth at the higher calcium level.

TABLE 12: Potassium, calcium, and magnesium contents of soybeans as influenced by variable amounts of potassium and calcium.

(Data for 50 plants. Crop 1, Series A)

Culture No.	Treatments		Potassium		Calcium		Magnesium	
	Potassium M. E.	Calcium M.E.	Per Cent	Total Mgms.	Per Cent	Total Mgms.	Per Cent	Total Mgms.
1	5	10	1.83	234.2	0.43	55.3	0.386	59.4
2	10	10	2.07	408.6	0.29	56.3	0.285	55.6
3	15	10	2.41	479.8	0.26	51.7	0.216	45.8
4	5	20	1.33	251.6	0.60	114.6	0.360	68.0
5	10	20	1.80	398.0	0.51	111.3	0.248	54.8
6	15	15	2.10	470.2	0.39	86.5	0.271	60.5

The total amount of potassium in the plant material was directly proportional to the amount of this element supplied, as is shown by the data of Table 12. The plants to which 5, 10, and 15 M. E. of potassium and 10 M. E. of calcium were available contained respectively 234.2, 408.6, and 479.8 mgms. of potassium. Plants supplied with the same amounts of potassium but with the higher level of calcium contained 251.6, 398.0 and 470.2 mgms. of potassium. The calcium supply had little influence on the absorption of potassium by the plants of the non-nodulated crop as is shown more clearly by the graph in Fig. 11.

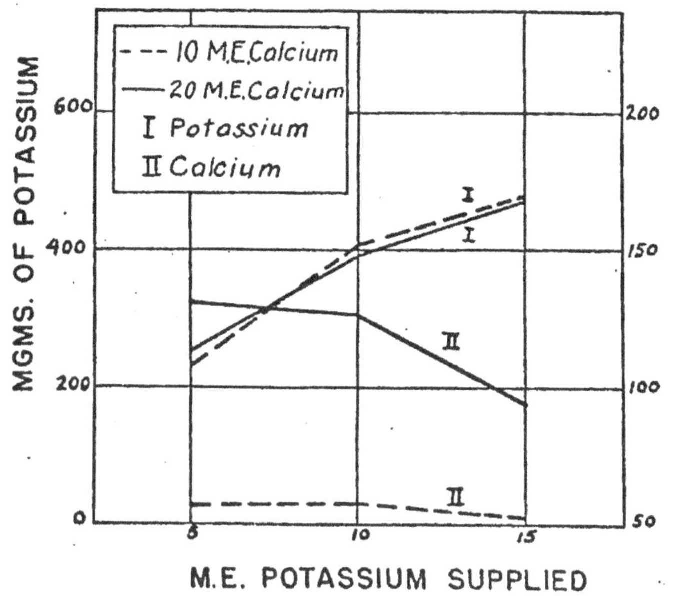

Fig. 11. Potassium and calcium contents of soybeans as influenced by variable amounts of these nutrient ions. (Crop 1, Series A).

Although the absorption of potassium in the nodulated and non- nodulated crops followed the same general trends, the non-nodulated plants of Series A utilized a lower percentage of the available potassium than did the plants of Series B which were inoculated, as a comparison of the data of Table 6 with that of Table 14 shows. In both crops, the efficiency of potassium absorption by the plants decreased with increments of potassium supplied. A greater calcium supply did not increase the efficiency of potassium utilization.

The ratios of nitrogen to potassium in both the nodulated and non-nodulated crops decreased as increments of potassium were supplied, as shown in Fig. 12. At the low potassium level, the ratios were increased by additions of calcium. In the plants supplied with 15 M. E. of potassium, there was little difference in the ratios of nitrogen to potassium. In general, the increments of potassium decreased the ratios of carbohydrate carbon to potassium in the plant tissues, as the graph of Fig. 13 shows. The larger carbohydrate content of the non-nodulated plants is indicated by their wider ratios of carbon to potassium, as well as by their increased ratios of carbon to nitrogen.

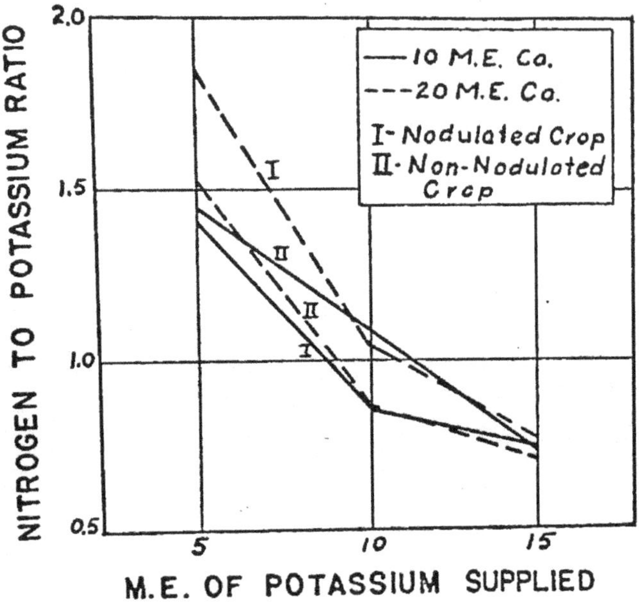

Fig. 12. Ratios of nitrogen to potassium as influenced by nitrogen fixation and variable amounts of potassium and calcium.

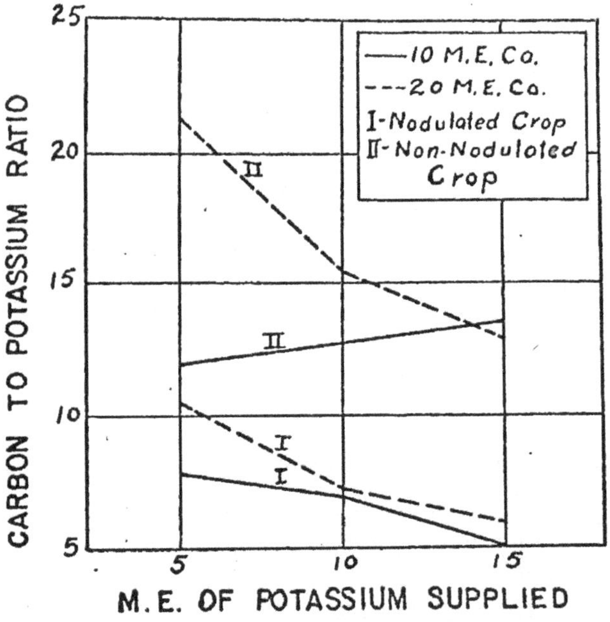

Fig. 13. Ratios of carbohydrate carbon to potassium as influenced by nitrogen fixation and variable amounts of potassium and calcium.

Calcium and Magnesium Contents—The calcium contents of the plants in this series were not significantly influenced by variable amounts of potassium. Although the concentration of calcium declined with additions of potassium, the decrease may be attributed to the increased weights of plant material. Cultures treated with 5, 10, and 15 M. E. of potassium and 10 M. E. of calcium contained a rather constant amount of calcium—namely 55.3, 56.3, and 51.7 mgms. respectively. The cultures supplied with 5 and 10 M. E. of potassium and 20 M. E. of calcium contained respectively 114.6 and 111.3 mgms. of calcium; and culture 6 supplied with 15 M. E. of potassium and 15 M. E. of calcium contained 86.5 mgms. of calcium.

The utilization by the plants of this series of the available calcium was quite constant, ranging from 24.4 to 27.8 per cent. The absorption of calcium by the plants of the nodulated crop was slightly less efficient than that of this crop which did not fix nitrogen.

At both levels of calcium, the concentration of magnesium in the plant tissue decreased as the potassium supply was increased. According to the data of Table 12, the magnesium content expressed as percentage was 0.386, 0.285, and 0.216 per cent respectively in cultures treated with 5, 10, and 15 M. E. of potassium and 10 M. E. of calcium; and 0.360, 0.248, and 0.271 per cent in cultures supplied with the same amounts of potassium but with the higher levels of calcium. The concentration of magnesium was not significantly influenced by the calcium level, a different situation than in soybeans manifesting leguminous behaviors.

The amount of magnesium in fifty plants decreased as increments of potassium were available, as shown by the data of Table 12. The magnesium contents of cultures supplied with 5, 10, and 15 M. E. of potassium and 10 M. E. of calcium were respectively 59.4, 55.6, and 45.8 mgms. For cultures treated with the same amounts of potassium but with the higher calcium levels the values were 68.0, 54.8, and 60.5 mgms. As was true in the case of the nodulated crop, calcium brought about an increased total uptake of magnesium.

The efficiency of magnesium absorption decreased as the supply of potassium increased, except in culture 6. The percentages of the available magnesium taken by the plants which fixed nitrogen were approximately the same as those utilized by the non-nodulated crop.

Phosphorus Contents—The concentrations of phosphorus in this crop were found to be very similar to those in the crop of soybeans which was inoculated. In general, the phosphorus contents as percentages of dry weight decreased with increments of potassium, as is shown in Table 13. The cultures treated with 5, 10, and 15 M. E. of potassium and 10 M. E. of calcium contained respectively 0.420, 0.343, and 0.297 per cent of phosphorus. The concentrations were respectively 0.324, 0.276, and 0.289 per cent in cultures supplied with the same amounts of potassium at the higher calcium levels. Again, since increased growth resulted from the greater supply of calcium, this nutrient element brought about a decrease in the concentration of phosphorus through increased growth.

TABLE 13: Barium, phosphorus, and silicon contents of soybeans in relation to variable amounts of potassium and calcium.

(Data for 50 plants. Crop 1, Series A)

Culture No.	Treatments		Barium		Phosphorus		Silicon	
	Potassium M. E.	Calcium M. E.	Per Cent	Total Mgms.	Per Cent	Total Mgms.	Per Cent	Total Mgms.
1	5	10	1.09	140.9	0.420	54.0	2.28	293.2
2	10	10	0.47	96.0	0.343	64.6	1.24	245.8
3	15	10	0.26	51.7	0.297	59.2	1.10	219.6
4	5	20	0.50	95.2	0.324	61.2	1.13	214.1
5	10	20	0.17	37.2	0.276	60.8	0.95	209.4
6	15	15	0.15	32.3	0.289	64.5	0.91	193.9

Cultures treated with 5, 10, and 15 M. E. of potassium and 10 M. E. of calcium contained a total of 54.0, 64.6, and 59.2 mgms. of phosphorus; while those supplied with the same increments of potassium coupled with 20 M. E. of calcium contained 61.2, 60.8, and 64.5 mgms. of phosphorus respectively. From the data of Table 15, it may be said that the total phosphorus consumed by non-nodulated soybeans was increased slightly by increments of both potassium and calcium.

TABLE 14: Consumption of potassium, calcium and magnesium as influenced by variable amounts of potassium and calcium.

Culture No.	Potassium			Calcium			Magnesium		
	Supplied Mgms.	Taken		Supplied Mgms.	Taken		Supplied Mgms.	Taken	
		Mgms.	Per Cent		Mgms.	Per Cent		Mgms.	Per Cent
1	363.5	234.2	64.6	212.1	55.3	26.0	132.1	59.4	44.9
2	559	408.6	73.2	212.1	56.3	26.5	132.1	55.6	42.1
3	754.5	479.8	63.7	212.1	51.7	24.4	132.1	45.8	34.8
4	363.5	251.6	69.2	412.1	114.6	27.8	132.1	68.0	51.4
5	559	398.0	71.2	412.1	111.3	27.1	132.1	54.8	41.5
6	754.5	470.2	62.4	312.1	86.5	27.7	132.1	60.5	45.8

Seed contained 168 mgms. Potassium, 12.1 mgms. Calcium, and 8.9 mgms. Magnesium.

TABLE 15: Consumption of barium, phosphorus, and nitrogen in relation to the total amounts supplied.*

(Data for 50 plants. Crop 1, Series A)

Culture No.	Barium			Phosphorus			Nitrogen**		
	Supplied Mgms.	Taken		Supplied Mgms.	Taken		Supplied Mgms.	Taken	
		Mgms.	Per Cent		Mgms.	Per Cent		Mgms.	Per Cent
1	1717.0	140.9	8.2	113.7	54.0	47.5	346	341.2	98.5
2	1373.6	96.0	6.9	113.7	64.6	56.7	346	-	-
3	1030.2	51.7	5.1	113.7	59.2	52.2	346	347.8	100.2
4	1030.2	95.2	9.2	113.7	61.2	53.8	346	384.5	111.0
5	686.8	37.2	5.4	113.7	60.8	53.6	346	341.2	98.5
6	686.8	32.3	4.7	113.7	64.5	56.7	346	334.2	97.2

*Fifty seeds contained 0.00 mgms. Barium, 36.2 mgms. Phosphorus, and 346 mgms. Nitrogen.
**This crop did not fix nitrogen.

Barium and Silicon Contents—As in the nodulated crop of soybeans, so the barium content of this crop expressed either as percentage of dry weight or as total amount in fifty plants was found to bear a direct relation to the amount of barium supplied. This is shown in Table 13. The consumption by the plants of the available barium was likewise very similar in the nodulated and non-nodulated crops as a comparison of the data of Tables 8 and 15 shows. From the results obtained, it seems that barium, being a cation, occupies a position in the ionic atmosphere of each colloidal clay particle according to its charge and effective size, and further that its absorption by plant roots is a matter of ordinary physico-chemical behavior.

The silicon contents of the plants declined regularly with increments of both potassium and calcium. The concentrations of silicon in cultures treated with 5, 10, and 15 M. E. of potassium and 10 M. E. of calcium were respectively 2.28, 1.24, and 1.10 per cent. The cultures supplied with the same amounts of potassium but with the higher levels of calcium contained 1.13, 0.95, and 0.91 per cent of silicon. A larger calcium supply brought about a significant decrease in the concentration of silicon particularly at the lower potassium levels. As increments of potassium were offered at the lower calcium level, the plants contained a total of 293.2, 245.8, and 219.6 mgms. of silicon respectively; and at the higher calcium levels there were 214.1, 209.4, and 193.9 mgms. of silicon. As was the case in Series B, the absorption of silicon by the plants depended upon the level of nutrient cations available to the plants rather than upon the base saturation of the adsorbing complex, which was nearly constant in all cultures.

Ratios of Plant Constituents

Calculations of the ratios existing between various plant constituents show some interesting contrasts between the nodulated and the non-nodulated soybeans. The ratios vary according to the increasing potassium combined with each of the calcium levels, as shown by the calculations assembled in Table 16. In most cases, the ratios decrease, in others they increase, as the potassium increases.

TABLE 16: Ratios of plant constituents

Crop 1, Series B (Nodulated)

Culture No.	Treatments		Ratios				
	K	Ca	N / Ca	K/ Ca	N / K	C / K	C* / N
1	5	10	10.1	7.2	1.40	8.2	5.8
2	10	10	8.0	9.4	0.85	6.6	7.8
3	15	10	8.4	11.3	0.74	5.0	6.8
4	5	20	5.2	3.1	1.85	10.4	5.6
5	10	20	4.9	4.7	1.04	6.7	6.4
6	15	20	4.5	5.9	0.77	6.0	7.8

Crop 1, Series A (Non-nodulated)							
1	5	10	6.2	4.2	1.45	12.0	8.5
2	10	10	-	-	-	-	-
3	15	10	3.8	5.3	0.73	13.4	18.6
4	5	20	3.3	2.2	1.52	21.3	13.9
5	10	20	3.1	3.6	0.86	15.9	18.0
6	15	15	3.8	5.4	0.71	12.7	18.5
*Carbohydrate carbon							

That protein production per unit of calcium was greater in the nodulated crops is suggested by its higher ratios of nitrogen to calcium. The higher ratios of potassium to calcium in the nodulated soybeans suggest a more liberal absorption of potassium from the soil induced by a higher concentration of nitrogen in the roots, and an adequate calcium supply on the clay. The crops differed only in that the nodulated crops, due to the activity of the legume bacteria, presented an additional supply of nitrogen. Although the colloidal-chemical conditions of the clay of the two sets of media were the same, the more proteinaceous roots of the nodulated cultures were able to take more potassium. The stimulating effect of nitrogen upon the absorption of potassium was pointed out by Breazeale (9), who concluded that the influence of the nitrogen is exerted within the plant tissues and is possibly related in some way with the process of absorption. Additional evidence of the effect of the nitrogen level within the plant tissues upon the intake of potassium is afforded by a comparison of the nitrogen to potassium ratios which are approximately the same at each level of potassium. The greater nitrogen content of the nodulated crop was balanced by an increased absorption of potassium. Although potassium is commonly associated with carbohydrate production, nitrogen fixation was not overshadowed by carbohydrate formation occasioned by the entrance into the plant of additional potassium along with the calcium.

In the nodulated crop, the lower ratios of carbohydrate carbon to nitrogen show that carbohydrate production per unit of potassium taken into the plant is lower with the occurrence of nitrogen fixation than when legume bacteria are absent. When nodulated, the same plant species functions more as a protein former than as a carbohydrate producer, yet consumes a greater amount of potassium. In soybeans that are not fixing nitrogen and where less potassium is absorbed in relation to the calcium, the formation of carbohydrates significantly dominates nitrogen synthesis, as the high ratios of carbon to nitrogen show. The carbon content of the non-nodulated plants is high in relation to the consumption of both potassium and nitrogen.

Discussion

The data obtained in this study point to the importance of both potassium and calcium in the growth and the nitrogen fixation of soybeans. From the fact that the concentration of nitrogen in the plant tissue decreased with increments of potassium at both levels of calcium, it appears that potassium has a greater influence than nitrogen or calcium on carbohydrate formation. The increasing ratios of carbohydrate carbon to nitrogen as additions of potassium were offered leads to the same conclusion.

An available potassium supply may be said to be one of the requirements for protein synthesis, inasmuch as it is rather well established that potassium is essential for the production and translocation of carbohydrates (10, 22, 25), and for the enzymatic induction of nitrates in the formation of amino acids (13). That higher levels of potassium stimulate carbohydrate production in excess of that required for meristematic activity and protein synthesis is shown by the increasing amounts of the carbohydrate fractions contained in the plant materials. In plants supplied with adequate potassium and abundant nitrogen the carbohydrates failed to accumulate, not because of a lack of these nutrients but probably because carbohydrates were used in large quantities in protein synthesis. The low sugar content of the nodulated crop suggests that such is the case. Nodulated soybeans suggest themselves as better forages because they have the energy value of non-legumes, and in addition a higher concentration of protein and minerals.

If a large amount of potassium is available in relation to nitrogen and calcium, then the plant material will be characterized by high carbohydrate and low protein content; which suggests the reason for the "woody" character of plants growing on soils relatively high in potassium. The ecological significance of this relationship has been pointed out by Albrecht (3).

The importance of calcium in nitrogen fixation has been emphasized repeatedly (1, 4, 21). The ability of the plant to fix nitrogen is directly related to increased calcium absorption. The role of calcium appears to be more directly connected with nitrogen fixation as indicated by the higher nitrogen and slightly lower carbohydrate concentrations in the plants grown at the higher calcium level.

Although some calculations presented by Albrecht (6) show that plants apparently secure more calcium than is adsorbed on the surface areas of colloidal clay particles with which the roots come into immediate contact, we may nevertheless reason that greater root surface should result in an increased amount of root-soil contact and, therefore, a larger intake of nutrient ions from the colloidal complex. In this study, increments of potassium brought about increased root weight and greater root development. If we assume that increased root weight is a measure of more extensive root surface, then the larger amount of surface area for contact absorption very probably facilitates the entrance of more calcium. The close agreement between the calculated and actual values of Table 17, which, taking the low treatment as a basis, shows the increased amount of calcium taken by the plants in proportion to the increased root weight for the nodulated crop, indicates that such a supposition is not without some foundation. The presence of an adequate potassium supply in the soil might be an important factor in the ability of the legume plant to take in sufficient calcium for the fixation of nitrogen.

The relative influence of either potassium or calcium appears to depend, at least to some extent, upon the available supply of the other. When the potassium supply was restricted, a higher calcium level greatly increased the growth and nitrogen fixation by the plants. As larger amounts of potassium were available, however, the effect of the greater calcium supply was not so marked, although the influence of a larger calcium supply was still significant. Legumes consume large amounts of potassium but they require so much more calcium that a low ratio of potassium to calcium is necessary for nitrogen fixation, as shown by the greater nitrogen content of the soybeans to which the higher level of calcium was available.

TABLE 17: Actual and calculated increased consumptions of calcium in relation to greater root weight as induced by increments of potassium.			
Culture No.	Increased Root Weight over Low Treatment (gms.)	Increased Consumption of Calcium by Plants (mgms.)	
		Calculated	Actual
1	-	-	-
2	–0.40	-	-
3	0.795	13.5	15.5
4	-	-	-
5	0.195	7.4	6.0
6	0.367	11.3	9.6

The data obtained in this investigation as well as those reported by Ferguson (14) show a greater production of dry weight by non-nodulated than by nodulated soybeans, although the initial nutrient levels of the cultures were the same in both series. This points out the fallacy in judging legume activity, particularly nitrogen fixation, in terms of tonnage alone. It is of interest, however, that the mineral concentrations in the non-nodulated plants were lower even if the totals were not very different. Apparently the abundant nitrogen in the nodulated crop resulted in an increased utilization of the carbohydrates for protein synthesis so that less vegetative growth occurred. In the plants adequately supplied with potassium and calcium and actively fixing nitrogen, the ratios of the carbohydrate carbon to the nitrogen were significantly lower than in the non-nodulated plants.

The accumulation of nitrates in low potassium plants was first reported by Stocklasa (33), and more recent studies (14, 34) suggest the loss of nitrogen from the plant to the soil under certain conditions. It seems possible that a restricted supply of potassium as well as nitrogen, although adequate for meristematic development, may result in increased growth of plants which are low in protein because of the lack of constituents essential for the synthesis of a large amount of protein; a portion of which becomes the "storage protein." If not utilized in the formation of more complex stable compounds, the nitrogen might conceivably be lost to the soil; particularly during maturation, with the consequent impairment of the nitrogen value of the forage.

That the concentration of the nutrient ions on the colloidal complex is a primary factor in deciding the amounts that move into the plants is indicated by the results obtained in this study. As increments of exchangeable potassium were offered more potassium was absorbed by the plants. The higher level of calcium likewise induced an increased intake of calcium. Since less nutrients on the clay meant more ionic barium on the clay and also more in the plants, it appears that the absorption of barium is regulated more by physico-chemical relations than by physiological factors associated with the roots.

However, the absorption of ions by the plants cannot be related entirely to physico-chemical relationships. The intake of silicon seemed to be limited when an adequate supply of nutrient cations was available; but when low levels of nutrient cations were offered, more silicon moved into the plants from a more or less constant supply. The data show that increased calcium absorption was brought about by increments of potassium, and the presence of a greater nitrogen supply as a result of nitrogen fixation was associated with an increased potassium content of the plant tissue. It was found that additions of calcium enhanced while those of potassium hindered the activity of the plants in utilizing magnesium. These results indicate that the intake of nutrient ions is apparently influenced by both physico-chemical and physiological behaviors, and point to the possibility of numerous interrelationships between ions. As we build up the concept of these interrelationships, might we not in time establish many others not yet recognized or anticipated, but which will aid greatly in the unravelling of the complexities of plant nutrition?

A comparison of some ratios of plant constituents in nodulated and non-nodulated soybeans which were given the same levels of all nutrients except nitrogen shows the decided influence of active legume bacteria upon the mineral consumption by plants. The mere rearranging of the amounts of cations on the soil substrate does not change a legume to a non-legume. In addition to containing a greater amount of calcium and nitrogen, the inoculated legume is characterized by a higher potassium content. Potash deficiencies as well as the lack of lime and phosphorus may be responsible for many legume troubles, as, for example, the failure of medium red clover which is known to have a higher potassium requirement than other legumes.

References Cited

1. Albrecht, Wm. A. *Calcium and Hydrogen Ion Concentration in the Growth and Inoculation of Soybeans*. Jour. Amer. Soc. Agron. 24: 793–806. 1932.
2. Albrecht, Wm. A. *Some Soil Factors in Nitrogen Fixation by Legumes*. Trans. Third Com. Internat. Soc. Sci. A: 71–84. 1939.
3. Albrecht, Wm. A. *Potassium-Phosphorus Relations as a Possible Factor in Ecological Array of Plants*. Jour. Amer. Soc. Agron. 32: 411–418. 1940.
4. Albrecht, Wm. A. and Davis, F. L. *The Physiological Importance of Calcium in Legumes Inoculation*. Bot. Gaz. 88: 310–321. 1929.
5. Albrecht, Wm. A., Graham, E. R. and Ferguson, C. E. *Plant Growth and the Breakdown of Inorganic Soil Colloids*. Soil Sci. 47: 455–458. 1938.
6. Albrecht, Wm. A., Graham, E. R. and Shepard, H. R. *Surface Relationships of Roots and Colloidal Clay in Plant Nutrition*. Amer. Jour. Bot. 29: 210–213. 1942.
7. Albrecht, Wm. A. and Schroeder, R. A. *Colloidal Clay Cultures for Refined Control of Nutritional Experiments with Vegetables*. Proc. Amer. Soc. Hort. Sci. 37: 689–696. 1939.
8. Bradfield, R. *The Chemical Nature of Colloidal Clay*. Mo. Agr. Exp. Sta. Res. Bul. 60. 1923.
9. Brazeale, J. F. *The Effect of One Element of Plant Food upon the Absorption by Plants of Another Element*. Univ. of Ari. Exp. Sta. Tech. Bul. 19. 1928.
10. Burwell, R. C. *Effect of Certain Deficiencies on Nitrogen Metabolism of Plants*. Bot. Gaz. 82: 320–328. 1926.
11. Crampton, E. W. and Maynard, L. A. *The Relation of Cellulose and Lignin to the Nutritive Value of Animal Feeds*. Jour. Nutrition 15: 383. 1938.
12. Dickson, J. G. *The Value of Certain Nutrient Elements in the Development of the Oats Plant*. Amer. Jour. Bot. 5: 301–325. 1918.
13. Eckerson, Sophia A. *Conditions Affecting Nitrate Reduction by Plants*. Contribution Boyce Thompson Inst. 4: 119–130. 1932.
14. Ferguson, C. E. and Albrecht, Wm. A. *Nitrogen Fixation and Soil Fertility Exhaustion by Soybeans under Different Levels of Potassium*. Mo. Agr. Exp. Sta. Res. Bul. 330. 1941.

15. Fonder, J. F. *Variations in Potassium Content of Alfalfa Due to Stage of Growth and Soil Type and the Relationship of Potassium and Calcium in Plants Grown upon Different Soil Types*. Jour. Amer. Soc. Agron. 21: 732–750. 1929.
16. Gieseking, J. E., Snider, H. J., and Getz, C. A. *Destruction of Organic Matter in Plant Material by the Use of Nitric and Perchloric Acids*. Ind. and Eng. Chem., Anal. Ed., 7: 185. 1935.
17. Graham, E. R. *Magnesium as a Factor in Nitrogen Fixation by Soybeans*. Mo. Agr. Exp. Sta. Res. Bul. 288. 1938.
18. Harris, H. C. *A Comparison of Potassium Permanganate and Ceric Sulfate for the Oxidation of Cobaltinitrite in the Estimation of Potassium in Potassium Chloride Solution and in Ammonium Acetate Soil-Extracts*. Soil Sci. 40: 301–309. 1936.
19. Heinze, P. H. and Murneek, A. E. *Comparative Accuracy and Efficiency on Determination of Carbohydrates in Plant Material*. Mo. Agr. Exp. Sta. Res. Bul. 314. 1940.
20. Hoagland, D. R. and Martin, J. C. *Absorption of Potassium by Plants in Relation to Replaceable, Nonreplaceable, and Soil Solution Potassium*. Soil Sci. 36: 1–33. 1933.
21. Horner, G. M. *Relation of the Degree of Base Saturation of Colloidal Clay by Calcium to the Growth, Nodulation, and Composition of Soybeans*. Mo. Agr. Exp. Sta. Res. Bul. 232. 1936.
22. Janssen, G. and Bartholomew, R. P. *The Translocation of Potassium in Tomato Plants and its Relation to their Carbohydrate and Nitrogen Distribution*. Jour. Agr. Res. 38: 447–465. 1929.
23. Jenny, H. and Cowan, E. W. *The Utilization of Adsorbed Ions by Plants*. Sci. 77: 394–396. 1933.
24. Kolthoff, T. M. and Sandell, E. B. *Textbook of Quantitative Inorganic Analysis*. Macmillan Company. New York. 1937.
25. Kornfeld, A. *Die Blattfleckenkrankheit der Soya-eine Kalimangel-Erscheinung. Zeitsehr.* Pflanzenernahr., Dung. u. Bodenk. 32: 201–221. 1933.
26. Loehwing, W. F. *Calcium, Potassium, and Iron Balance in Certain Crop Plants in Relation to their Metabolism*. Plant Physiol. 3: 261–275. 1928.
27. Loew, O. Zur *Physiologie von Kalium und Kalzium*. Ernahr. der Pflanze 30: 141–144. 1934.
28. Marshall, C. E. Layer *Lattice and Base Exchange Clays*. Ztschr. Krist. A: 91: 433–449. 1935.

29. Murneek, A. E. and Heinze, P. H. *Speed and Accuracy in Determination of Total Nitrogen*. Mo. Agr. Exp. Sta. Bul. 261. 1937.
30. Nightingale, G. T., Schermerhorn, L. G., and Robbins, W. R. *Some Effects of Potassium Deficiency on the Histological Structure and Nitrogenous and Carbon Constituents of Plants*. N. J. Agr. Exp. Sta. Bul. 499. 1930.
31. *Official and Tentative Methods of Analysis of the A. O. A. C.* Washington, D. C. 1935.
32. Reed, H. S. *The Value of Certain Elements to the Plant Cell*. Ann. Bot. 21: 501–543. 1907.
33. Stocklasa, J. *Ist das Kalium an der Eiweiss-synthese in der Pflanzenzelle Beteiligt?* Biochem. Ztschr. 73: 107–161. 1916.
34. Virtanen, A. I., Von Hausen, S., and Laine, T. *Investigations on Root Nodule Bacteria of Legume Plants. XIX, Influence of Various Factors on the Excretion of Nitrogenous Compounds from the Nodule*. Jour. of Agr. Sci. 27: 332–348. 1937.
35. Wall, M. E. *The Role of Potassium in Plants: I. Effect of Varying Amounts of Potassium on Nitrogenous, Carbohydrate, and Mineral Metabolism in the Tomato Plant*. Soil. Sci. 47: 143–161. 1939.
36. Weiszmann, H. *Uber den Einfluss des Kaliums auf die Entwicklung der Pflanzen und ihren Morphologischen und Anatomischen Bau, bei Besonderer Berucksichtigung der Landwirtschaftlichen Kulturpflanzen*. Zeitschr. Pflanzendung. 2:1–79. 1923.

CHAPTER 13

Nitrogen Fixation and Soil Fertility Exhaustion by Soybeans Under Different Levels of Potassium[1]

Abstract

A study was made of the influence of potassium on nitrogen fixation, plant composition, and fertility depletion by soybeans. An electrodialyzed colloidal clay of known composition was used as a carrier of this cation in combination with constant amounts of calcium, magnesium, phosphorus, and in some instances, sulfur. Barium was used to vary reciprocally with potassium, thereby maintaining a constant degree of saturation.

In nodulated cultures nitrogen fixation increased as the available potassium increased, when accompanied by constant levels of the other elements. The ratio of nitrogen to carbohydrate was approximately constant. Plants which were non-nodulated were very low in nitrogen and did not fix nitrogen, but, instead, lost to the soil an average of 23.1 per cent of the initial seed supply. The carbohydrate contents of these plants increased tremendously both in percentage and in total with potassium treatment. Crops grown on soils depleted in fertility were nodulated but failed to fix nitrogen. But they produced increasing amounts of carbohydrates as more potassium was present in the plants. Higher potassium treatments caused a decrease in the total amounts of magnesium taken. All the media tended to approach the same degree of saturation of cations other than barium when cropped successively. The efficiency of phosphorus removal by soybeans was increased greatly by potassium treatment. Under certain conditions

[1] *Carl E. Ferguson and Wm. A. Albrecht, "Nitrogen Fixation and Soil Fertility Exhaustion by Soybeans Under Different Levels of Potassium," Research Bulletin 330 (Missouri Agricultural Experiment Station, May 1941).*

the crop contents of potassium, nitrogen, and phosphorus were less than those present in the planted seed. The cations calcium and magnesium were always present in the harvested crop in excess of the amounts supplied by the seed.

The study assembles many aspects of soil fertility in their relation to nitrogen fixation by soybeans.

Introduction

An available supply of potassium has long been recognized as a necessity for successful plant growth. Although its specific functions are not completely understood, it is fairly well established that, among other activities, potassium plays an essential role in carbohydrate metabolism, particularly in that phase involving sugar and starch formation. Plants may differ widely in their content of potassium. Some species absorb and retain potassium in such large quantities that they have been designated as "potassium plants." Invariably plants of this nature are high in carbohydrate material.

Calcium is also an indispensable cation in plant nutrition. Here too, as in the case of potassium, plant species differ considerably in their calcium requirements and in their calcium contents. The legumes, for example, are especially exacting in their requirement for this nutrient. This is well demonstrated by the fact that liming of soils low in calcium is an essential practice in order to ensure healthy nodule formation, nitrogen fixation, and high quality protein production.

For the purpose of simplicity, protein production in plants can be regarded as the result of the union of two types of compounds: the first, carbohydrates or a substance for which carbohydrate is the precursor, and, the second, available nitrogen. Since potassium is intimately concerned in the formation of carbohydrate material, and since calcium is known to be effective in legume behavior, it appears that the organic chemical constituents of plants may be influenced greatly by the amount of available potassium in relation to other cations.

Such a consideration is of particular significance in legume production since legumes are grown primarily because of their capacity to fix nitrogen and to produce high quality protein. The question might well be raised whether legumes grown under the widely varying soil conditions of kind and amount of nutrients, of ratios between nutrients, and of different degrees of activity of legume bacteria, are necessarily high in nitrogen and protein simply because they are members of the family Leguminosae. Is it not possible that soil fertility conditions play a significant part in determining the chemical composition of the organic as well as the inorganic fractions of plant material?

The study reported herein was undertaken in order to determine more accurately the influence on growth and plant composition by a varying supply of potassium when delivered along with constant amounts of the other fertility items, viz. phosphorus, calcium, and magnesium. Coincident with the study of plant composition, attention was also directed toward the qualitative and quantitative changes brought about in the nutrient supply on the substrate as a result of cropping. It was in the hope of learning more about the relations of potassium to other nutrient cations and their behavior within the soil and the legume plants for nitrogen fixation, that the following study was undertaken.

Historical

The widespread occurrence of potassium in plant and animal tissue has been recognized for some time. In 1905 McCallum (29) investigated a wide range of animal and vegetable tissues. He pointed out the great variation of tissues in potassium content.

Many studies have been made which show the increase in potassium content of plants with potassium fertilization. Among these studies are those by Janssen and Bartholomew (17), Johnston and Hoagland (21), and Hart (13). In many of these studies, "luxury" consumption of the element was indicated at the higher potassium levels.

Fundamental researches on the function of potassium in plant metabolism have revealed that the physiological activities in which potassium plays a significant role probably include: (1) the formation and transformation of carbohydrates; (2) the relation to nitrogen; (3) the catalytic effects and; (4) the formation and storage of oils.

That carbohydrates are formed normally only when the potassium supply is adequate was suggested by Nobbe in 1871 (31). After studying a variety of green plants, Reed (37) concluded that potassium is essential for starch formation. More recently Janssen and Bartholomew (18) (19) have investigated the relationship between the potassium supply and carbohydrate production. For their work with cowpeas they reported a close correlation between the per cent potassium and the total weight of starch and sugars in the plant. Janssen and Bartholomew, and Hart (13) working with sugar cane, found that plants low in potassium were high in reducing and total sugars but lower in starch than plants adequately supplied with potassium.

The lowering of carbohydrate production by potash deficiency is probably closely related to the effect of potash on respiration and carbon dioxide assimilation. Briggs (6) showed in 1922 that respiration could not go on normally in plants with an insufficient supply of potassium. Alten and Goeze (4), in a carefully controlled series of experiments reported that respiration of wheat plants, within limits, was dependent on the amount of potassium supplied. The potassium concentration and carbon dioxide assimilation showed a close correlation. Plants with potassium deficiency aged rapidly and were unable to assimilate carbon dioxide. Maiwald (26) showed that potassium fertilization of plants growing under low light intensities resulted in a considerable increase in the efficiency of carbon dioxide assimilation and dry matter production.

That potassium might possibly be an essential mineral element in protein formation was first suggested by Stocklasa (38). He considered that with adequate supplies of carbohydrate and other materials, protein can be formed either in the presence or absence of potassium. In the dark, protein formation occurred only in the presence of potassium. Loew (24) also suggested that potassium is concerned in protein formation in cells, pointing out as evidence, the large quantities of potassium in seeds containing abundant protein reserves. Although potassium has been reported to be essential for protein formation, yet many workers have reported a decrease in protein concentration by potassium treatment. Janssen and Bartholomew (19) found the percentage of nitrogen to be highest in cowpeas grown in media deficient in this element. Rauterberg and Loofmann (36) grew oats and rye grass under increasing levels of potassium and with three levels of nitrogen. At all levels of nitrogen, the increasing potassium resulted in a declining percentage of total nitrogen. However, the total nitrogen harvested increased with potassium treatment.

The depressing effect of potassium on the feeding capacity of plants for other ions has been given considerable attention. Fonder (8) reported an inverse relation between calcium and potassium contents of alfalfa plants. Johnston and Hoagland (21) showed that with low supplies of potassium, tomatoes absorbed relatively more calcium, magnesium, and phosphorus. Drake and Scarseth (7) noted a lowering of the percentage of calcium and of magnesium in tobacco plants by fertilization with potassium. A chlorotic condition of the tobacco leaves was corrected by applications of magnesium sulfate.

The various studies reported have clearly demonstrated the importance of an adequate supply of potassium for normal plant performance. That it plays a role in carbohydrate metabolism has been recognized and clearly demonstrated. However, the effect of potassium on nitrogen metabolism and protein formation is rather obscure. Information on the influence of potassium on symbiotic nitrogen fixation by legumes is particularly inadequate. The studies suggest that potassium may serve in metabolic processes concerning both carbohydrate and nitrogen. A study of this question by supplying legume plants with variable supplies of potassium, and noting the changes in plant composition should indicate more clearly the influence of this element on nitrogen fixation and plant composition.

Methods and Materials: Preparation and Description of Hydrogen-Clay[2]

In order to simulate soil conditions as nearly as possible, and at the same time to provide a medium which could be brought under chemical scrutiny both at the outset and at the end of a cropping season, an electrodialyzed colloidal clay from the subsoil of the Putnam silt loam was used as a carrier for the nutrient cations. Media of this nature have been successfully used previously in similar studies (1), (2), (11). Colloidal clay, of which the chemical and mineralogical properties have been described by Bradfield (5) and Marshall (27) respectively, was separated from the subsoil of the Putnam silt loam. Particles with an effective diameter of less than .200 micron were fractionated by means of the Sharples centrifuge. H-clay was prepared from this fine fraction by electrodialysis, thereby freeing the clay of essentially all adsorbed ions other than hydrogen. The H-clay thus prepared had a pH of 3.6. Its acidic properties were those as shown by the curve of Fig. 1. The total exchange capacities of two samples used were determined and found to be 65 and 68 M E [milligram-equivalents] per hundred grams of clay.

The clay thus prepared is very similar in physical and chemical properties to that part of the soil exchange complex which, through its base exchange properties, is credited with a very dynamic role in the storage

[2] *Hydrogen-clay represents the acid clay after electrodialysis to remove the other cations and replace them by exchangeable hydrogen. It will be spoken of as H-clay.*

and release of nutrient elements to plants (33) (2). Its chemical composition is shown in Table 1. Although it contains a relatively abundant supply of potassium, magnesium, iron, and a small amount of calcium, previous work (3) has shown that during the course of a growing season, with the single exception of iron, no significant quantities of those elements within the clay crystal are taken up by plants.

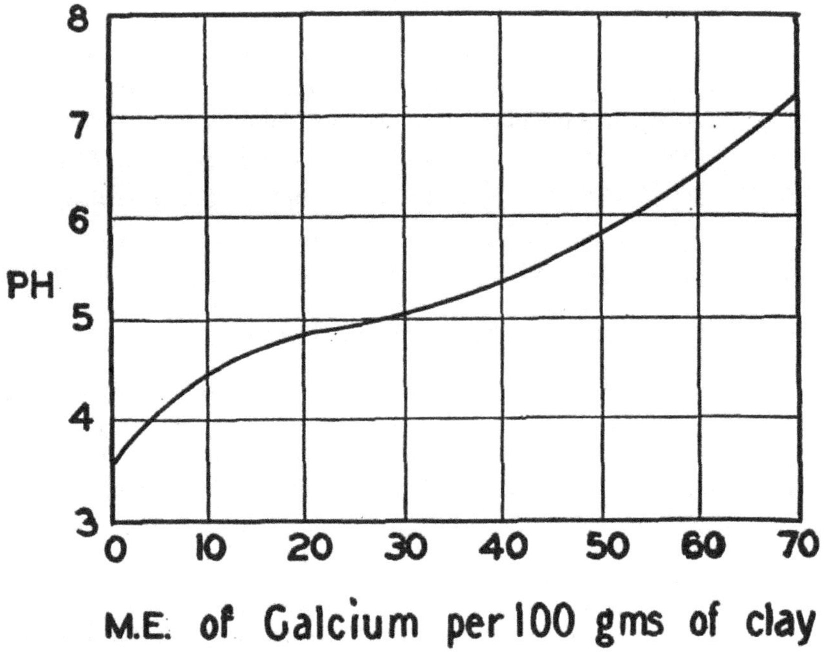

Fig. 1. Potentiometric titration curve of hydrogen clay by calcium hydroxide.

Carbon and nitrogen analyses of the electrodialyzed clay gave values of 1.56 and 0.157 per cent respectively. This carbon and this nitrogen are held in the colloidal humic matter. Since it represents the more stable colloidal humic residues formed during organic matter decay and retained against electrodialysis, it would seem doubtful whether appreciable quantities of this nitrogen could be brought into an available form during a short cropping period of five weeks. Experimental trials demonstrated the stability of this humic matter.

TABLE 1: Composition of electrodialized Putnam colloidal clay* (.2 micron fraction).	
Element	**Per Cent**
Si	23.0
Al	12.77
Fe	5.73
K	0.924
Mg	0.661
Na	0.422
Ca	0.177
Mn	0.0188
C	1.56**
N	0.157**

*According to C. E. Marshall (27)
**Analyses by author.

After growing a series of three crops of soybeans on a colloidal clay-sand culture the clay was separated, washed and electrodialyzed again. The carbon content of the clay was the same as that of the original clay, except for only a slight increase. The nitrogen values, however, actually increased, which could be accounted for only by a loss of nitrogen from the soybean crop to the soil. Thus the decomposition of the carbon and nitrogen of the humic residue in the clay surely could not have been contributing much to the crop growth.

Culture Preparation

The media were prepared by taking a sufficient amount of the H-clay to sorbe the desired quantity of bases and adding the nutrients as hydroxides whenever possible. Barium was used as a balancing cation reciprocating with potassium and thereby maintaining a constant degree of saturation of the clay. The suspensions were allowed to stand for several days after the addition of each of the cations. A definite sequence of additions was followed. This, in order of sequence was, barium hydroxide, magnesium oxide, phosphorus as mono-calcium acid phosphate, calcium hydroxide, and potassium hydroxide. In one series 0.5 M E of the magnesium, as magnesium sulfate, was added to each culture. Proper physical condition of the clay was obtained by the addition of 2000 grams of acid leached quartz sand to the clay suspensions and allowing the mixture to evaporate to optimum moisture.

Soybean plants were grown in two successive sets of media. The first group of plants is designated as Series A and the second, as Series B. Series B differed from Series A in that 0.5 M E of the magnesium was added as magnesium sulfate. Series A was used to produce one crop while series B served for three successive crops without nutrient addition beyond that given initially. Constant amounts of calcium, magnesium, and phosphorus were supplied to all the cultures. Potassium varied from 0 M E to 15 M E, while barium varied reciprocally to the potassium in each series. Treatments and arrangement of the series are schematically shown in Table 2.

Soybeans of the Virginia variety were germinated until the radicles were about 2 cms. long. They were then inoculated with Rhizobium japonicum and planted into the sand-colloidal clay mixture at the rate of fifty per pot. All cultures were in duplicate. Crop one of Series A was grown in the greenhouse during the Fall for a period of six weeks. Crops of Series B were each grown for five weeks. Crop one of Series B was uninoculated and grew outside under a partial shade during July. Crop two grew during October, and crop three was planted the following May. Artificial light was used to supplement daylight during the latter days of the growth period of the Fall crops.

TABLE 2: Nutrients added by titration to cultures of Series A and B.

Series A (Used for a single crop)						
Culture No.	Nutrient ions added					
	Ca	Mg	P	K	Ba	
	M E	M E	M E	M E	M E	
1	10	5	3.68	0	15	
2	10	5	3.68	5	10	
3	10	5	3.68	10	5	
4	10	5	3.68	15	0	

Series B (Used for Three Successive Crops)						
Culture No.	Nutrient ions added					
	Ca	Mg	P	S	K	Ba
	M E	M E	M E	M E	M E	M E
1	10	5	3.68	0.5	0	15
2	10	5	3.68	0.5	5	10
3	10	5	3.63	0.5	10	5
4	10	5	3.63	0.5	15	0

At the end of the growing period all crops were harvested about mid-day and placed in the autoclave at five pounds pressure for five minutes in order to stop any enzyme action. After enzyme inactivation they were dried to a constant weight in the oven at 60°C., weighed separately as to tops and roots, and then finely ground together.

Analytical Methods

Chemical analyses were made for the following constituents: nitrogen, calcium, magnesium, potassium, phosphorus, barium, iron, aluminum, silicon, reducing sugar, total sugar, starch, and hemicellulose. Portions of the finely ground samples were digested by the wet ashing method (9). The elements calcium, magnesium, phosphorus, potassium, iron, silicon, aluminum, and barium, were determined in the digested sample. Plant silica was separated from any contaminating material by the official methods (32). Barium was found both in the hydrochloric acid solution of the ash, and as insoluble barium sulfate in the residue from the silica separation. Methods described by Kolthoff and Sandell (22) were used for the determination of the inorganic elements. Separate portions of the plant material were taken for total nitrogen determinations by the Kjeldahl-Gunning-Arnold method as modified by Murneek and Heinze (30). Carbohydrate analyses were made according to the methods of Shaffer and Somogyi as modified by Heinze and Murneek (14).

Experimental Results: "Non-legume" Behavior. Crop One—Series B

The presentation of the experimental results is divided into the following three divisions: (1) non-legume as contrasted to legume behavior under different potassium treatments; (2) declining fertility levels as reflected in yields and crop composition; and (3) the extent of nutrient exhaustion in the soil through nutrient removal by successive croppings. The results reported herewith include the data for one crop of Series A and for three crops of Series B. Attention will first be given Crop one—Series B which was non-nodulated, and although a legume, performed in a manner entirely different from the commonly ascribed legume behavior. The influence of variable supplies of potassium on plant performances is given.

Variable Potassium in Relation to Growth—Although the crop was supplied only the inherent seed nitrogen, yet as is shown in Table 3, relatively good growth in terms of total dry matter yields was obtained. Cultures 1, 2, 3, and 4 produced crops which were 3.39, 4.41, 5.08, and 4.67 times heavier respectively than the weight of the planted seed. Increasing potassium gave corresponding increases in growth, with the exception of the highest potassium treatment, culture 4, which showed a decline below culture 3 in total crop weight. However, it may be noted that the root growth of this culture was largest of all. Growth was very vigorous until the fourth week at which time nitrogen deficiency symptoms were manifested by a cessation of growth, and by the development of a yellowish green color in the leaves. Plate 1 shows the plants at the time of harvest.

Plate 1. Soybean plants as influenced by increments of potassium. (Crop One—Series B. Cultures I, II, III, and IV had 0, 5, 10, and 15 M E respectively of potassium supplied.)

TABLE 3: Crops of soybeans in relation to variable potassium. (Data for 50 plants. Crop one—Series B.)								
Culture No.	Application of Potassium	Weight of 50 Plants			Ratio Tops	Increase of Tops Over No Potassium	Increase of Roots Over No Potassium	Ratio Crop Wt.
		Tops	Roots	Total				
	M E	gms.	gms.	gms.	Roots	%	%	Seed Wt.*
1	0	7.821	6.878	14.699	1.14	...	...	3.39
2	5	12.064	7.252	19.316	1.66	54.2	5.44	4.41
3	10	15.264	7.245	22.509	2.11	95.2	5.34	5.08
4	15	12.582	7.643	20.225	1.65	61.0	11.12	4.67

*Fifty seeds weighed 4.43 gms.

TABLE 4: Composition of plant material in relation to variable potassium. (Data for 50 plants. Crop one—Series B.)											
Culture No.	Application of Potassium	Potassium		Calcium		Phosphorus		Magnesium		Barium	
	M E	Per Cent	Total mgms.	Per Cent	Total mgms.	Per Cent	Total mgms.	Per Cent	Total mgms.	Per Cent	Total mgms.
1	0	0.73	111.2	0.44	66.7	0.196	29.3	0.34	51.8	1.44	216.3
2	5	1.32	258.5	0.41	80.4	0.196	38.2	0.28	55.6	0.93	181.4
3	10	1.68	379.4	0.41	92.2	0.203	45.7	0.25	56.3	0.42	94.5
4	15	2.18	450.3	0.43	88.7	0.214	44.1	0.23	43.6	0.00	0.00

TABLE 5: Consumption of potassium, calcium, and nitrogen in relation to the total amounts supplied. (Data for 50 plants. Crop one—Series B.)									
Culture No.	Potassium			Calcium			Nitrogen		
	Supplied* mgms.	Taken		Supplied* mgms.	Taken		Supplied* mgms.	Taken	
		mgms.	Per Cent**		mgms.	Per Cent**		mgms.	Per Cent**
1	178	111.2	62.4	208.1	66.7	32.0	318	235.4	–26.0
2	373.5	258.5	69.2	208.1	80.4	38.6	318	241.9	–23.8
3	569	379.4	66.7	208.1	92.2	44.3	318	261.0	–17.9
4	764.5	450.3	59.0	208.1	83.7	42.5	318	243.1	–23.6

*Seed contents were: Potassium 178 mgms., calcium 8.1 mgms., and nitrogen 318 mgms.
**The content of the plants as per cent of the total supplied by the seed and the titration is given. In case of the nitrogen, this percentage represents the increase over the initial nitrogen or per cent fixation.

Something of the morphological nature of the growth is shown by the ratios of tops to roots. With increasing potassium, the ratio of the tops to the roots widened considerably, increasing from a low value of 1.14 in culture 1 with no potassium to 2.11 in culture 3. Culture 4 with a ratio of 1.65 had a lower figure than culture 3 or a close approach to culture 2 at 1.66. Expressed as percentage increase of top growth over culture 1, the culture with treatments of 5, 10, and 15 M E of potassium gave values of 54.2, 95.2, and 61.0 respectively. Percentage increases in root growth were only of the order of 5.44, 5.34, and 11.12. The much larger increase in root growth of culture 4 is also shown by the corresponding culture of Series A crops one and two.

Potassium Content—The amount of potassium present in the plant tissue, expressed either as percentage concentration or total quantity, was in direct relation to the amount offered. The culture treated with 0, 5, 10 and 15 M E of potassium, as Table 4 shows contained 0.73, 1.32, 1.68, and 2.18 per cent of potassium respectively, or totals of 111.2, 258.5, 379.4, and 450.3 mgms. per culture. When the total potassium content of the crop is plotted as a function of the amount supplied, the graph of Fig. 2 approaches a straight line, curving but slightly with the higher treatments. The percentage of potassium taken from that supplied varied little with the higher treatments. The data of Table 5 show that of the total potassium supplied, which included that in the seed plus that titrated on the clay, the percentages taken were 62.4, 69.2, 66.7, and 59.0. It is very significant that culture 1 with no titrated potassium, although growing on a medium saturated with bases and approximately neutral at the outset, lost 37.6 per cent of its initial seed potassium, or a loss of 65.8 mgms. of an initial seed supply of 178 mgms. Yet this culture did not show growth irregularities commonly considered as typical potassium deficiency symptoms.

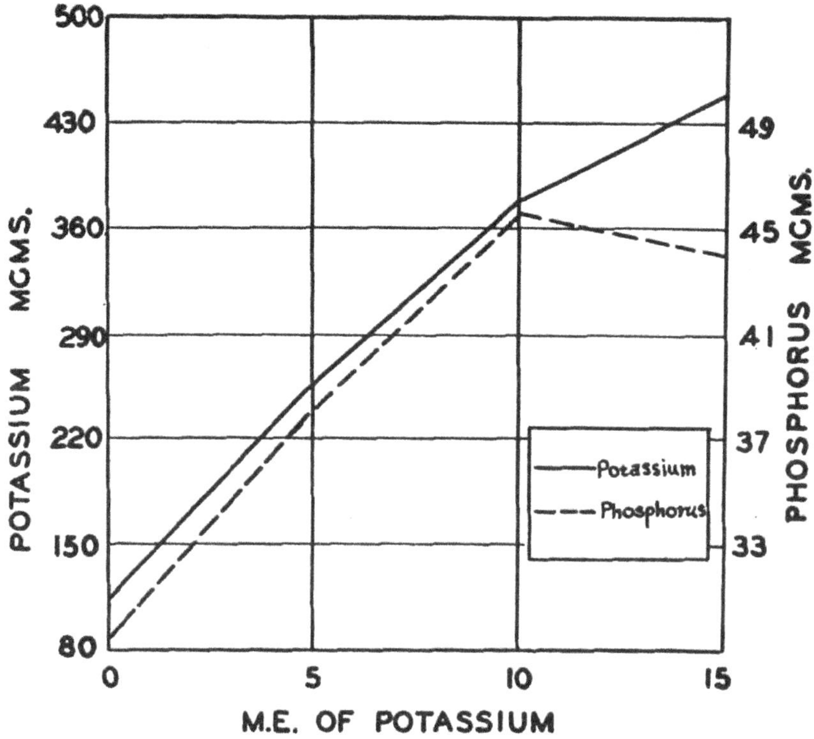

Fig. 2. Potassium and phosphorus contents of soybeans as influenced by potassium. (Crop One—Series B.)

Nitrogen Content—Since the crop was not inoculated with nitrogen fixing bacteria, and since nitrogen was not supplied along with the other nutrients, one would expect to find nitrogen in the plant tissue only to the extent that it occurred in the seed. But, even though this crop produced yields of dry matter ranging from 15.0 to 22.5 grams per culture, which was 3.39 to 5.08 times the weight of the planted seed, there was actually a negative nitrogen balance. All cultures in this crop contained less nitrogen than was in the seed from which they grew. The amount of this loss is given in Table 5 and represented graphically in Fig. 3. Increasing the potassium offered had little effect on this nitrogen loss. This loss is significant, since of the 318 mgms. of nitrogen supplied by the fifty seeds, culture 1 lost 83 mgms. or 26.0 per cent; culture 2, 76 mgms. or 23.8 per cent; culture 3, 57 mgms. or 17.9 per cent; and culture 4, 75 mgms. or 23.6 per cent. These values give a mean loss of 23.1 per cent of the seed nitrogen back to the soil. So severe was this loss that the nitrogen concentration of the harvested crop ranged from as low as 1.16 per cent to no higher than 1.57 per cent, as is shown in Table 6, on the basis of the dry forage for this legume crop. This percentage content in nitrogen suffered a decrease with the increase in applied potassium.

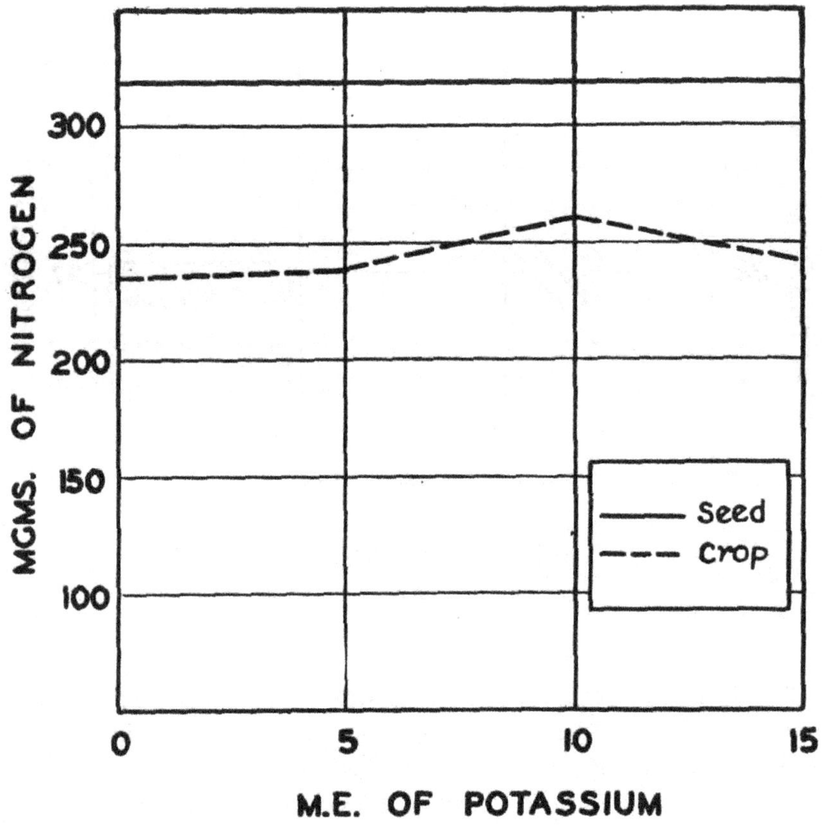

Fig. 3. Nitrogen losses by non-nodulated soybeans. (Crop One—Series B.)

TABLE 6: Nitrogen content in relation to variable potassium. (Data for 50 plants. Crop one—Series B.)					
Culture No.	Application of Potassium	Nitrogen in 50 Plants			Seed Nitrogen
	M E	Per Cent	Total mgms.	Fixed mgms.	Lost Per Cent
1	0	1.57	235.4	–82.6	26.0
2	5	1.24	241.9	–76.1	23.8
3	10	1.16	261.0	–57.0	17.9
4	15	1.18	243.0	–74.9	23.6

Significance of Nutrient Losses—The possibility of a reverse movement of nutrients, particularly from plants growing on highly adsorptive substrates has been recognized for some time. Lundegardh (25) suggested that there might be a competition between plant colloids and soil colloids for nutrient cations and that possibly some cations could be lost to the substrate. Albrecht (1) working with different Ca-H-clays found in some cases that the harvested crops had less of certain elements than the seed from which they grew.

Thomas (39) suggested that plants may lose potassium back to the soil after the formation of blossoms. Petrie (35) found that the total potassium decreased in plants after maturity was reached. He found no loss of calcium however. More recently Jenny (20) has shown that excised roots, although still undergoing respiration, lost potassium readily when immersed in certain kinds of clays. H-clay was most effective in removing potassium.

Since there was no loss of plant parts during the growth and harvesting of the crop it may be assumed that the nitrogen unaccounted for in the harvested plants, moved into the sand-colloidal clay substrate. The evidence for a movement of this kind is further substantiated by the analyses of the substrates which are given on pages 211–24.

It is of interest that cultures 1, 2, 3, and 4 developed from lots of seed weighing 4.43 grams per culture produced dry matter to the extent of 3.39, 4.41, 5.08, and 4.67 times their initial seed weight and at the same time suffered an average loss of 23.1 per cent of the initial seed nitrogen back to the soil. These increments in dry matter, regardless of the high nitrogen loss, are probably accounted for by the fact that the dry matter produced was largely carbohydrate in composition. Since it contains little or no nitrogen, only a minimum amount of nitrogen need be present for its production. Carbohydrate synthesis and storage were stimulated tremendously by the absorption and retention of increasing amounts of potassium.

These data emphasize the fallacy of judging the value and quality of hays solely on the basis of appearance or species, with no consideration of the soil conditions under which they were produced. It is possible to harvest a crop of soybeans containing less nitrogen than was contained in the planted seed.

The foregoing data suggest that a certain level of available potassium must be present on the clay to prevent a migration of seed potassium back to the soil, even though the clay may be completely saturated and approximately neutral at the outset. The relatively good growth made by plants losing potassium is possible because of the seed content of potassium, which is high enough to supply the plants' potassium needs for a relatively long period of time after germination.

Carbohydrate Content—The carbohydrates varied in close parallel with the growth, according to Fig. 4. Both reducing sugars and total sugars decreased in percentage with increasing potassium supplied. Cultures supplied with 0, 5, 10, and 15 M E of adsorbed potassium contained 2.16, 1.92, 1.85, and 0.94 per cent reducing sugars respectively as is shown by Table 7. The relatively low level of sugars may be explained by the low rate of photosynthetic activity at the time of harvest, which was an effect of the extreme nitrogen shortage. Nevertheless, carbohydrate was stored in abundant quantities during the early growth period as is shown by the very high starch content of all cultures. The effect of potassium in bringing about an increase in the percentage and in the total starch is very evident. Cultures 1 to 4 show starch contents of 8.39, 10.65, 17.35, and 15.52 per cent or totals of 1.263, 2.078, 3.913, and 3.210 grams respectively. Expressed as percentage increase in total starch over culture 1, cultures 2, 3, and 4 gave values of 64.5, 210, and 154 per cent, respectively. The hemicellulose remained almost constant in percentage concentration. The data show that in plants with low nitrogen but with increasing supplies of available potassium and constant amounts of other nutrient elements, the rates of carbohydrate synthesis and storage are increased tremendously.

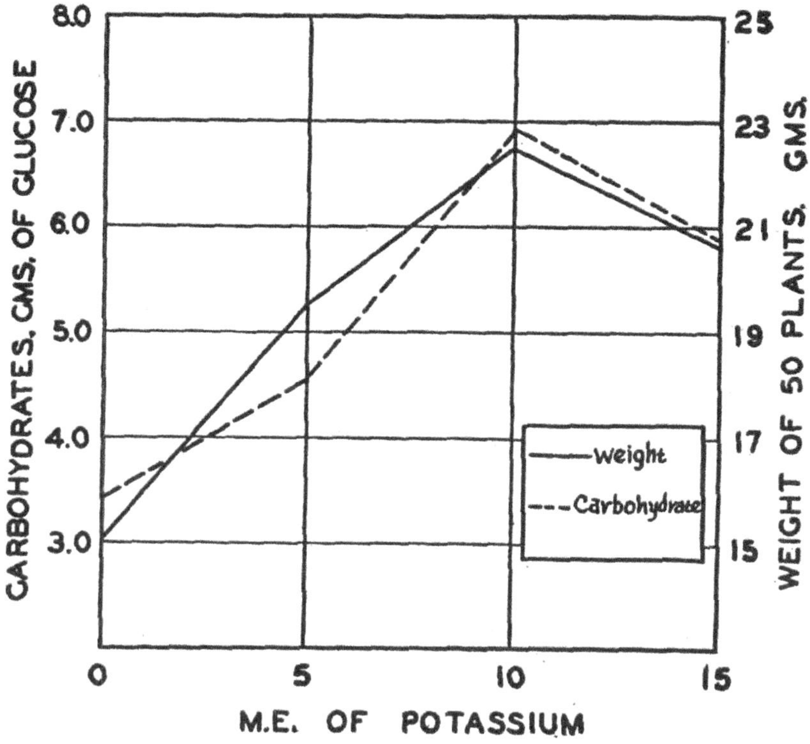

Fig. 4. Growth and total available carbohydrates, as influenced by variable potassium. (Crop One—Series B.)

TABLE 7: Carbohydrate content of plant material as influenced by potassium. (Data for 50 plants. Crop one—Series B.)

Culture No.	Sugar				Starch		Hemicellulose		% Reducing Sugar of the Total Sugar
	Reducing		Total						
	Per Cent	Total mgms.	Per Cent	Total mgms.	Per Cent	Total mgms.	Per Cent	Total mgms.	
1	2.16	324	3.89	582	8.39	1263	10.40	1563	55.7
2	1.92	374	2.68	523	10.63	2078	9.94	1940	71.6
3	1.85	418	2.42	544	17.35	3913	11.11	2499	76.5
4	0.94	195	1.31	271	15.52	3210	11.59	2396	71.8

Culture No.	Phosphorus			Magnesium			Barium		
	Supplied* mgms.	Taken		Supplied* mgms.	Taken		Supplied* mgms.	Taken	
		mgms.	Per Cent**		mgms.	Per Cent**		mgms.	Per Cent**
1	70	29.3	41.9	71.6	51.8	72.3	1030.2	216.3	21.0
2	70	38.2	54.6	71.6	55.6	77.6	686.8	181.4	26.4
3	70	45.7	65.3	71.6	56.3	78.6	343.4	94.5	27.6
4	70	44.1	63.0	71.6	43.6	60.8	...	...	...

TABLE 8: Consumption of phosphorus, magnesium, and barium in relation to the total amounts supplied. (Data for 50 plants. Crop one—Series B.)

*Seed contents were: Phosphorus 32 mgms., magnesium 10.8 mgms., and barium 0.00 mgms.
**The content of the plants as per cent of the total supplied by the seed and the titration is given.

Cation and Phosphorus Intake—The calcium concentration in the plant tissue remained fairly constant with increasing potassium. Since more growth was obtained with potassium increments, the total calcium also increased. (See Table 4). Cultures 1 to 4 show total calcium contents of 66.7, 80.4, 92.2, and 88.7 mgms. respectively, which represent a consumption of 32.0, 38.6, 44.3, and 42.5 per cent of the 208.1 mgms. as total calcium supplied as shown in Table 5. Here again, as in all cases before, culture 4 failed to show an increase over the culture 3.

The percentage content of phosphorus of the plant material was almost constant as given in Table 4, but the total phosphorus harvested showed significant increases as more potassium was supplied with a single exception. Cultures supplied 0, 5, 10, and 15 M E of potassium contained 29.3, 38.2, 45.7, and 44.1 mgms. of phosphorus respectively, or 41.9, 54.6, 65.3, and 63.0 per cent of the constant supply of 70 mgms. as given in Table 8. Thus the total phosphorus consumption by soybeans was increased by the presence of increasing amounts of available potassium.

Unusually large concentrations of any nutrient cation in the plant tissue are ordinarily reflected by reciprocal concentrations of some other cation or cations. The amount of magnesium in the plant tissue, expressed as percentage of dry weight, decreased regularly and consistently with potassium treatment as shown in table 4. Cultures given treatments of 0, 5, 10, and 15 M E of potassium had magnesium contents of 0.34, 0.28, 0.25 and 0.23 per cent respectively. Since more growth was made by the potassium treated cultures, total magnesium, with the exception of culture 4, increased. It is significant to note that the quantities of magnesium taken by these cultures represent 72.3, 77.6, 78.6, and 60.8 per cent of the total supply of 71.6 mgms. offered as is shown by the data of Table 8. The efficiency of absorption and retention of magnesium by soybeans when supplied on a colloidal clay under these conditions, was of about the same magnitude as for potassium.

Barium was supplied as the reciprocal of the potassium. Interestingly, its amount in the plant material expressed either as per cent, of the dry weight or as total milligrams per culture was directly proportional to the amounts supplied. Although barium is regarded as a non-essential element for plant growth, yet considerable quantities were assimilated as is shown in Table 4. Cultures 1, 2, and 3 contained 1.44, 0.93, and 0.42 per cent respectively. Culture 4 had no barium supplied either in the seed or clay. The percentage taken of the total barium supplied was the lowest of all the cations. From a total supply of 1030, 687, and 343 mgms. only 21.0, 26.4, and 27.6 per cent respectively, were removed in the crop tissue as shown in Table 8.

These data suggest that barium is taken in relatively large amounts with apparently no significant injury to the crop growth. As an inert cation to occupy part of the adsorption capacity and yet give no plant injury it seems to have served very well.

Legume Behavior. Crop One—Series A

Unlike the crop described previously as a non-legume behavior, crop one of Series A was inoculated with active legume bacteria and fixed nitrogen. The influence of variable supplies of potassium on growth, nitrogen fixation, ion absorption, and plant composition are of interest.

Variable Potassium in Relation to Growth—Potassium treatment brought about increased growth, which was approximately proportional to the amount of potassium supplied. Plate 2 shows the plants at the time of harvest. Cultures supplied 0, 5, 10, and 15 M E of potassium produced 11.880, 14.219, 15.036, and 16.007 gms. of plant material as is given in Table 9 and shown graphically by Fig. 5. There is an interesting relationship between roots and tops. The cultures from 1 to 4 inclusively had top: root ratios of 1.81, 2.14, 2.32, and 2.35 respectively. The much greater per cent increase in top growth than root growth is also shown in Table 9 expressed as percentage increase in both top and root growth over culture 1. Culture 4 with the highest level of potassium, as in Series B, showed a marked increase in root growth over culture 3. These results indicate that there is no strict proportionality between top growth and the extent of root absorbing surface.

Plate 2. Soybean plants as influenced by increments of potassium. (Crop One—Series A. Cultures I, II, III, and IV were supplied 0, 5, 10, and 15 M E respectively of potassium.)

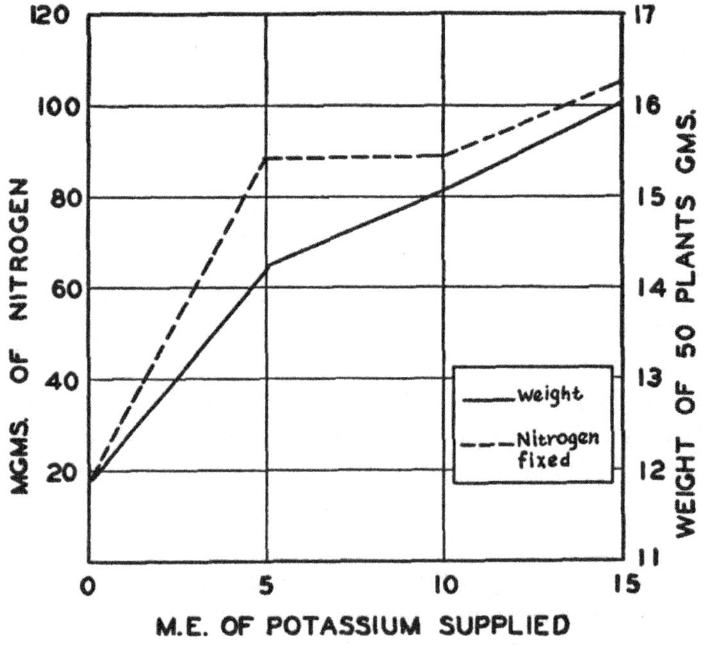

Fig. 5. Growth and nitrogen fixation as influenced by variable potassium. (Crop One—Series A.)

TABLE 9: Crops of soybeans in relation to variable potassium. (Data for 50 plants. Crop one—Series A.)							
Culture No.	Application of Potassium M E	Weight of 50 Plants			Increase Over No Treatment %		Ratio Tops Roots
		Tops gms.	Roots gms.	Total gms.	Tops	Roots	
1	0	7.641	4.238	11.880	...	...	1.81
2	5	9.688	4.531	14.219	26.8	6.93	2.14
3	10	10.506	4.530	15.036	37.5	6.90	2.32
4	15	11.220	4.787	16.007	46.9	12.95	2.35

Potassium Content—Potassium intake in this nitrogen fixing crop followed the same general behavior as that of the "non-legume," crop one of Series B. The data presented in Table 10 show that the potassium, expressed either as per cent or total amount, was in direct relation to the amount supplied. Cultures given 0, 5, 10, and 15 M E of potassium contained 88.5, 223.7, 369.8, and 471.4 mgms. respectively at the time of harvest. This high potassium intake had a depressing effect on the absorption of some other elements. This crop was very similar to crop one—Series B in percentage consumption of the total amount supplied, since 49.6, 59.8, 65.0, and 61.6 per cent of the total amounts supplied were taken. The high mobility of potassium was again shown by culture 1 which had no titrated potassium. Although the medium was completely saturated and approximately neutral at the outset, about fifty per cent of the initial seed supply of potassium was lost to the clay. This suggests again that a certain concentration of available potassium must be present on the clay for soybeans if they are to retain even their initial seed potassium during a cropping season.

TABLE 10: Nitrogen fixation by soybeans in relation to variable potassium. (Data for 50 plants. Crop one—Series A.)							
Culture No.	Application of Potassium M E	Number of Nodules per 50 Plants	Amount in 50 Plants				
			Potassium		Nitrogen		
			Per Cent	Total mgms.	Per Cent	Total mgms.	Fixed mgms.
1	0	453	0.74	88.5	2.81	335.7	17.7
2	5	438	1.69	223.7	2.86	405.5	88.5
3	10	618	2.51	369.8	2.70	406.9	88.9
4	15	629	2.94	471.4	2.64	423.3	105.3
Fifty seeds contained 318 mgms. of nitrogen and 178 mgms. of potassium.							

Nitrogen Fixation and Nodule Formation—Nitrogen fixation and nodule formation increased from the lowest to the highest level of potassium as is shown by the data of Table 10. Fig. 5 shows that increased growth was accompanied by a corresponding increase in nitrogen fixation. The percentage of total nitrogen declined, with the exception of culture 2, as the potassium supply increased. The favorable effect of treatment on bacterial infection and nodule formation is also indicated by the large numbers of nodules which were present on the cultures. The potassium treatment increased the carbohydrate production markedly in the "non-legume" performer; but in this case of the legume performer, although the potassium in the tissue increased from 0.74 per cent in culture 1 to 2.94 per cent in culture 4 or a four fold increase, yet nodule formation, nitrogen fixation, and protein production were enhanced considerably.

Carbohydrate Content—Since this crop was actively functioning in nitrogen fixation and protein production, consequently the organic composition of the plant tissue differed from that of crop one—Series B, which was performing as a non-legume. The reducing sugar values, according to the data of Table 11, are very similar to those of the non-nitrogen fixing crop, namely 1.91, 1.91, 1.04, and 0.96 per cent respectively for cultures 1, 2, 3, and 4. The decrease in this constituent is to be expected if one of the functions of potassium is to accelerate the conversion of sugar to starch. Values for total sugar are somewhat erratic but show a downward trend with increasing potassium added to the soil. The starch contents were much lower in magnitude than in the preceding crop. These values were 3.37, 2.64, 3.62, and 3.25 per cent for cultures supplied with increasing potassium in that order. That this crop was functioning in a manner entirely different from that of crop one—Series B is further shown by the nearly constant ratio of available carbohydrate carbon to nitrogen in Fig. 8.

Cation and Phosphorus Intake—The calcium concentrations in the plant tissue, according to the data of Table 12, were approximately constant, varying only from 0.43 to 0.48 per cent. The total calcium absorbed by the plants increased with increasing quantities of potassium supplied, so that cultures given 0, 5, 10, and 15 M E took 57.6, 60.2, 66.1, and 70.4 mgms. respectively of calcium. These values represent 27.6, 28.9, 31.8, and 33.8 per cent of the constant total supply of 208.1 mgm. of calcium. That calcium should have decreased only slightly in percentage and yet have shown an increase in total quantity in the plant material in spite of the tremendous saturation of the tissue with potassium can probably be accounted for by the fact that magnesium showed regular and consistent decrements both in percentage concentration and in total amount metabolized as the potassium treatments increased. The increase in tissue production by the treated cultures also served to raise the total amount of calcium utilized.

With increasing potassium supply, magnesium decreased regularly both in percentage and in total amount as is indicated by Table 12 and shown graphically in Fig. 6. The plant tissue from cultures supplied 0, 5, 10, and 15 M E of potassium contained 0.473, 0.369, 0.297, and 0.236 per cent, of magnesium or totals of 56.2, 50.8, 44.3, and 37.7 mgms. respectively. In spite of the declining magnesium utilization, the growth and the nitrogen fixation both continued to increase as more potassium was taken. The percentage absorption of the initial magnesium supplied by the seed and titration, ranged from a high of 78.5 per cent to a low of 52.7 per cent, as is shown in Table 13.

TABLE 11: Carbohydrate content of plant material as influenced by variable potassium. (Data for 50 plants. Crop one—Series A.)

Culture No.	Sugar				Starch		Hemicellulose		% Reducing Sugar of Total
	Reducing		Total						
	Per Cent	Total mgms.	Per Cent	Total mgms.	Per Cent	Total mgms.	Per Cent	Total mgms.	
1	1.91	227	4.06	483	3.37	401	14.69	1.745	47.0
2	1.91	255	4.09	566	2.64	360	13.90	1.811	45.0
3	1.04	157	2.94	444	3.62	544	12.93	1.943	35.1
4	0.96	154	3.01	482	3.25	520	9.32	1.492	31.9

TABLE 12: Composition of plant material in relation to variable potassium. (Data for 50 plants. Crop one—Series A.)

Culture No.	Application of Potassium M E	Potassium		Calcium		Magnesium		Phosphorus		Barium	
		Per Cent	Total mgms.	Per Cent	Total mgms.	Per Cent	Total mgms.	Per Cent	Total mgms.	Per Cent	Total mgms.
1	0	0.74	88.5	0.48	57.6	0.473	56.2	0.229	27.9	1.02	121.6
2	5	1.69	223.7	0.43	60.2	0.369	50.8	0.206	27.6	0.82	114.9
3	10	2.51	369.8	0.43	66.1	0.297	44.3	0.213	31.2	0.55	83.0
4	15	2.94	471.1	0.44	70.4	0.236	37.7	0.195	31.2	...	...

Fifty soybean seeds contained 178 mgms. of potassium, 10.8 mgms. of magnesium, 8.1 mgms. of calcium, and 23.8 mgms. of phosphorus.

TABLE 13: Consumption of phosphorus, magnesium and barium in relation to the total amounts supplied. (Data for 50 plants. Crop one—Series A.)

Culture No.	Phosphorus			Magnesium			Barium		
	Supplied* mgms.	Taken		Supplied* mgms.	Taken		Supplied* mgms.	Taken	
		mgms.	Per Cent**		mgms.	Per Cent**		mgms.	Per Cent**
1	61.8	27.9	45.2	71.6	56.2	78.5	1030.2	121.6	11.8
2	61.8	27.6	44.7	71.6	50.8	71.0	686.8	114.9	16.7
3	61.8	31.2	50.5	71.6	44.3	61.9	343.4	83.0	24.3
4	61.8	31.2	50.5	71.6	37.7	52.7	...	...	...

*Seed contents were: Phosphorus 23.8 mgms., magnesium 10.8 mgms., and barium 0.00 mgms.
**The content of the plants as per cent of total supplied by the seed and the titration is given.

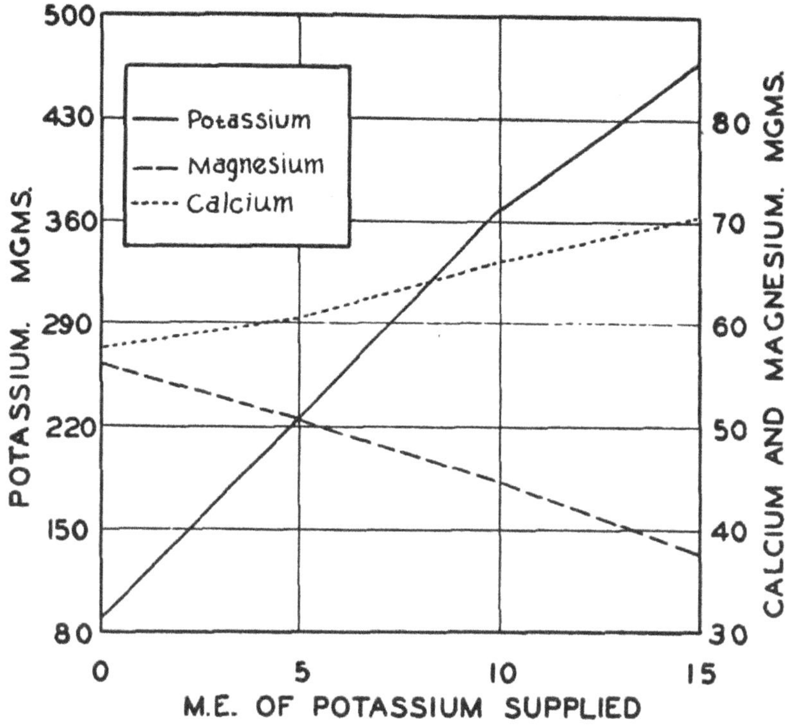

Fig. 6. Potassium, magnesium, and calcium contents of soybeans as influenced by variable potassium. (Crop one—Series A.)

According to the data of Table 12, phosphorus was present in the plant tissue in amounts ranging from 0.229 per cent in culture 1 to 0.195 per cent in culture 4. However, the total phosphorus increased slightly. The utilization of phosphorus varied between 45 and 50 per cent of that initially supplied by seed and titration, as is shown in Table 13. This is lower for this anion than that for potassium, and higher than that for calcium, both cations, as given in Table 14.

Effects on Crop Growth and Composition by Reductions in Soil Fertility. (Crop two—Series B).

In order to learn of the effects of declining fertility levels on growth, nitrogen fixation, nutrient absorption, and plant composition the media of crop one—Series B were used to produce two additional crops, without additional treatment. Both of these additional crops were inoculated with legume bacteria. There was a fallow period of 6 months following the growth of the second crop and before the planting of the third. The fallow sand-clay cultures were watered regularly to prevent the clay from developing undesirable physical properties.

Growth—Throughout the entire period of growth, the second crop showed no irregularities associated with nitrogen deficiencies. But, at the end of the third week, symptoms which are commonly associated with potassium deficiency began to appear. A light yellowing around the marginal areas of the older leaves, immediately followed by the development of brownish areas of wrinkled necrotic tissue was the usual course. These symptoms may be observed in Plate 3. Analyses of the tissue showed that all the cultures were very low in potassium. The relatively low harvest of dry matter per culture, as given in Table 15, was a result of the lower level of nutrients offered.

TABLE 14: Consumption of potassium, calcium, and nitrogen in relation to the total amounts supplied. (Data for 50 plants. Crop one—Series A.)

Culture No.	Potassium			Calcium			Nitrogen		
	Supplied* mgms.	Taken		Supplied* mgms.	Taken		Supplied* mgms.	Taken	
		mgms.	Per Cent**		mgms.	Per Cent**		mgms.	Per Cent**
1	178	88.5	49.6	208.1	57.6	27.6	318	335.7	5.56
2	373.5	223.7	59.8	208.1	60.2	28.9	318	406.5	27.8
3	569	369.8	65.0	208.1	66.1	31.8	318	406.9	27.9
4	764.5	471.1	61.6	208.1	70.4	33.8	318	423.3	33.1

*Seed contents were: Potassium 178 mgms., calcium 8.1 mgms., and nitrogen 318 mgms.

**The content of the plants as per cent of the total supplied by the seed and the titration is given. In case of the nitrogen, this percentage represents the increase over the initial nitrogen in the seed or per cent fixation.

TABLE 15: Growth of soybeans in relation to declining fertility. (Data for 50 plants. Crop two—Series B.)

Culture No.	Application of Potassium M E	Weight of 50 Plants			Ratio Tops Roots	Nodules No.	Per Cent Increase Over No Treatment	
		Tops gms.	Roots gms.	Total gms.			Tops	Roots
1	0	6.933	3.177	10.485	2.20	268	...	...
2	5	7.095	3.073	10.377	2.31	242	1.46	—3.28
3	10	7.287	3.408	11.375	2.14	263	4.21	7.30
4	15	8.327	4.175	13.602	2.00	304	19.10	31.40

Plate 3. Soybean plants as influenced by declining fertility. A, B, and C represent crops one, two, and three respectively of Series B.

The ratios of tops to roots varied from a low of 2.00 in culture 4 to a high of only 2.31 in culture 2, as is shown in Table 15. These values are much higher than the corresponding ratios of crop one of the same series, which were 1.11, 1.66, 2.11, and 1.65 for increasing potassium treatments in that order. However, they are of the same magnitude as those of crop one—Series A. Top-root ratios seem to be closely correlated with the nitrogen metabolism, since the crop which did not fix nitrogen made extensive root growth while crop one of Series A fixed nitrogen and made a relatively larger proportion of top to root growth. All cultures of this second crop were well nodulated, and there was no great variation in number of nodules per culture.

Potassium Content—The potassium contents of the harvested cultures increased with the increasing potassium treatment, but yet the harvested potassium represented a loss from the seed supply in all cultures, except culture 4. The potassium contents of cultures 1, 2, 3, and 4 according to the data of Table 16 were 91.2, 105.8, 157.0, and 195.9 mgms. respectively. The losses from the initial seed potassium by cultures 1, 2, and 3 represent 49.7, 41.4, and 13.3 per cent respectively of the 181 mgms. supplied by the seed. In the preceding crop, cultures 2 and 3 absorbed and retained the titrated potassium in abundant quantities, but the second crop on these cultures did not take up any of the remaining titrated potassium. On the converse, it contributed part of the initial seed potassium to the soil. Culture 1 was again "fertilized" by seed potassium.

TABLE 16: Nitrogen fixation by soybeans as influenced by declining fertility. (Data for 50 plants. Crop two—Series B.)

Culture No.	Application of Potassium M E	Crop Weight gms.	Nitrogen			Potassium	
			Per Cent	Total mgms.	Fixed mgms.	Per Cent	Total mgms.
1	0	10.485	3.31	346	28	0.87	91.2
2	5	10.377	3.34	346	28	1.02	105.8
3	10	11.375	3.33	378	60	1.38	157.0
4	15	13.602	3.18	433	115	1.44	195.9

Fifty seeds contained 318 mgms. of nitrogen and 181 mgms. of potassium.

Nitrogen Fixation and Nodule Formation—Nitrogen fixation and nodule formation increased from the lowest to the highest levels of potassium, as is shown by the data of Table 16 and graphically by Fig. 7. Cultures originally supplied with 0, 5, 10, and 15 M E of potassium fixed 28, 28, 60, and 115 mgms. of nitrogen respectively. Thus, liberal quantities of nodules were formed in cultures 1, 2, and 3 and nitrogen was fixed, in spite of the fact that potassium was being lost back to the clay in quantities sufficient to show characteristic potassium deficiency symptoms in the growing plants. The amount of seed potassium was sufficient to support plant growth for a period long enough for nodulation and the nitrogen-fixing process to take place.

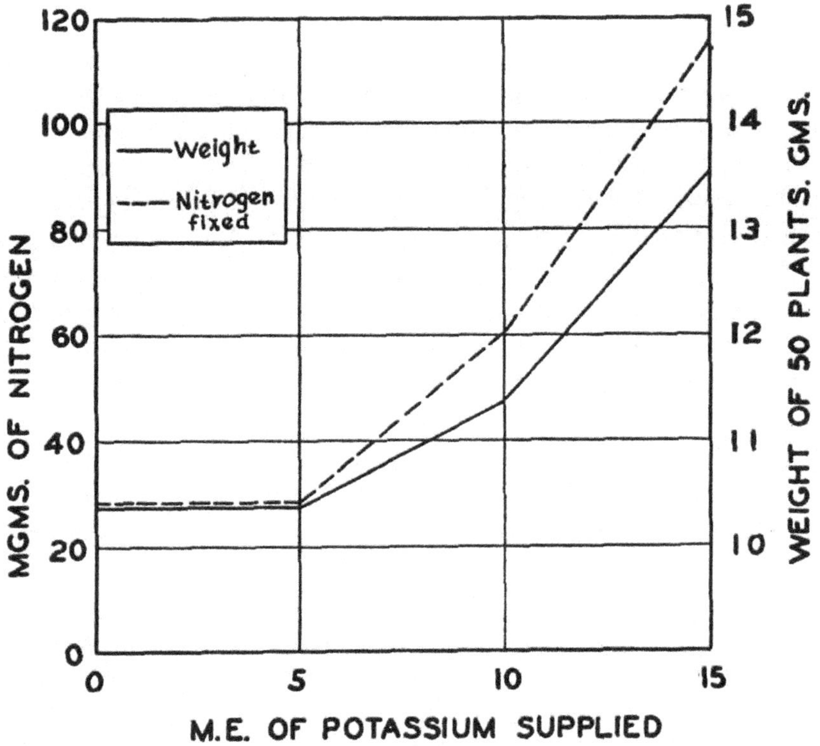

Fig. 7. Growth and nitrogen fixation as influenced by variable potassium under declining levels of fertility. (Crop two—Series B.)

Absorption of Bases and Phosphorus—The calcium contents of the second crop, according to the data of Table 17, were higher in percentage than in the preceding crop. This probably can be accounted for by the lower potassium absorption and considerably smaller yields of dry matter. The amounts of calcium removed and of nitrogen fixed showed a close correlation. The low percentage concentration and total amount of magnesium was the result of the very high magnesium consumption by the preceding crop, since from 60.8 to 78.6 per cent of the titrated magnesium had been removed by crop one. Cultures 1, 2, 3, and 4 contained only 21.2, 19.8, 21.7, and 30.0 mgms. of magnesium respectively. Culture 4 took up the largest quantity of magnesium since, as is shown in Table 8, this culture had the largest amount of magnesium remaining after crop one.

The phosphorus contents showed the usual decline with lower fertility levels. The beneficial effect of the higher levels of potassium on phosphorus consumption was shown in this crop by the increase in phosphorus of cultures 3 and 4 over the initial seed supply, while cultures 1 and 2, with low levels of potassium, lost some initial seed phosphorus to the soil. They retained only 28.5 and 28.7 mgms. respectively of the initial seed supply of 32.8 mgms. Even phosphorus, which is usually found in complex chemical combinations in the plants, can be lost from the living plant when it is growing on a highly absorptive medium carrying a low supply of bases and a low degree of base saturation.

TABLE 17: Composition of plant material as influenced by declining fertility. (Data for 50 plants. Crop two—Series B.)

Culture No.	Application of Potassium M E	Potassium		Calcium		Magnesium		Phosphorus		Barium	
		Per Cent	Total mgms.	Per Cent	Total mgms.	Per Cent	Total mgms.	Per Cent	Total mgm	Per Cent	Total mgms.
1	0	0.87	91.2	0.64	67.1	0.20	21.2	0.272	28.5	1.09	114.8
2	5	1.02	105.8	0.54	56.0	0.19	19.8	0.277	28.7	0.71	73.7
3	10	1.38	157.0	0.53	60.2	0.19	21.7	0.290	33.2	0.32	36.9
4	15	1.44	195.9	0.57	78.2	0.22	30.0	0.263	35.7	...	...

Fifty soybean seeds contained 181 mgms. of potassium, 11.5 mgms. of magnesium. 8.9 mgms. of calcium and 32.8 mgms. of phosphorus.

TABLE 18: Carbohydrate content of plant material as influenced by declining fertility. (Data for 50 plants. Crop two—Series B.)									
Culture No.	Sugar				Starch		Hemicellulose		% Reducing Sugar of Total
	Reducing		Total						
	Per Cent	Total mgms.	Per Cent	Total mgms.	Per Cent	Total mgms.	Per Cent	Total mgms.	
1	6.80	711	8.68	907	4.21	440	10.77	1102	78.4
2	5.96	632	7.46	787	4.11	441	9.94	1031	79.8
3	5.03	570	8.60	978	4.09	465	12.71	1445	58.5
4	4.95	674	7.69	1046	5.19	705	12.73	1698	64.4

Carbohydrate Production—The percentage concentration of the reducing sugars decreased with the increasing residual potassium, as is shown in Table 18. The total sugars were considerably higher, both in concentration and in totals, than in the corresponding cultures of the preceding crop. This difference may be explained by the extreme nitrogen hunger of the first crop at the time of harvest. It is significant that the starch percentages and totals were relatively constant. The low starch contents are in sharp contrast to those of crop one of the same series, but show a close similarity to crop one of the nitrogen fixing series. This relationship is further shown by the performance of a third crop on this medium. Hemicellulose increased in total but varied somewhat in percentage concentration.

Significance of Fertility Levels as Revealed in Plant Performances—The fertility level at which crop two was performing was still sufficiently high to permit nitrogen fixation. It is significant that nitrogen fixation can occur in plants which show a net loss of potassium and of phosphorus at the time of harvest, as was the case for cultures 1 and 2.

The fact that potassium absorption and retention occurred only in culture 4 may be explained by the lower level of bases and lower degree of potassium saturation. There is also the possibility of fixation of potassium into a non-exchangeable or unavailable form. Tests made on the soil after growing a third crop showed that a considerable quantity of the potassium originally titrated into the soil could not be recovered from the soil as exchangeable potassium by the customary exchange reagents.

Effects on Crop Growth and Composition by Further Reductions in Soil Fertility. (Crop Three—Series B).

The legume behavior under very low fertility levels was represented by crop three, since there were no additional treatments added to the soil. The various plant performances as influenced by declining fertility levels are decidedly suggestive of the soil's importance.

Growth—The amount of growth made by crop three is given in Table 19. A residual effect on yields by the potassium originally provided as increasing levels is still in evidence. Cultures 1, 2, 3, and 4 produced yields of 11.938, 12.364, 14.504, and 12.420 grams respectively. These yields are significant when one considers that all cultures showed definite symptoms of mineral deficiencies, particularly such as are usually associated with a lack of potassium. In addition, symptoms usually ascribed to a lack of magnesium and of phosphorus were exhibited by some of the cultures. Plate 3, row C, shows the appearance of the crop at the time of harvest.

The root growth in relation to the growth of tops was markedly influenced by the low level of the fertility elements supplied. A greater portion of the dry matter produced was carried by the tops. The ratios of tops to roots gave values of 2.52, 2.53, 2.59, and 3.04 respectively. The preceding crop had ratios considerably smaller than these suggesting that with declining fertility a smaller relative amount of the crop is left as root organic matter in the soil for the succeeding crop.

TABLE 19: Growth and nodulation of soybeans as influenced by declining fertility. (Data for 50 plants. Crop three—Series B.)

Culture No.	Application of Potassium M E	Weight of 50 Plants			Ratio Tops Roots	Nodules No.
		Tops gms.	Roots gms.	Total gms.		
1	0	8.536	3.402	11.938	2.52	103
2	5	8.860	3.504	12.364	2.53	106
3	10	10.471	4.033	14.504	2.59	50
4	15	9.337	3.083	12.420	3.04	1

Potassium Content—Potassium intake, according to Table 20, showed a negative balance with the single exception of culture 4, which increased slightly. The amounts of potassium harvested in cultures 1 and 2 exceed those harvested in the corresponding cultures of the previous crop. This is not wholly unexpected. There was actually more potassium supplied to the cultures of crop three, than to those of crop two since some of the seed potassium of the previous crop had gone to the soil and had "fertilized" it with potassium. This fertilization was not of inconsiderable magnitude as cultures 1, 2, and 3 of the preceding crop lost 90, 75, and 24 mgms. of potassium respectively, back to the soil.

TABLE 20: Nitrogen fixation and potassium content as influenced by declining fertility. (Data for 50 plants. Crop three—Series B.)

Culture No.	Application of Potassium M E	Crop Weight gms.	Nitrogen			Potassium	
			Per Cent	Total mgms.	Fixed mgms.	Per Cent	Total mgms.
1	0	11.938	2.55	305	−13	1.12	133.8
2	5	12.364	2.68	331	13	1.30	160.6
3	10	14.504	2.21	320	2	1.05	151.8
4	15	12.420	2.70	330	12	1.50	193.8

Fifty seeds contained 318 mgms. of nitrogen and 181 mgms. of potassium.

Nitrogen Fixation and Nodule Formation—Although cultures 1, 2, 3, and 4 were nodulated to the extent of 103, 106, 50, and 1 nodule respectively per culture, none demonstrated any significant gain in nitrogen in the form of nitrogen fixation as is shown in Table 20. The failure of this crop to fix nitrogen in any significant amounts may be explained by the critically low levels of calcium, magnesium, and phosphorus in the cultures. All these elements are required in liberal quantities if nitrogen fixation is to proceed normally. Potassium could not have been the limiting factor for nitrogen fixation, since cultures 1 and 2 of the preceding crop contained less potassium, but yet produced greater numbers of nodules and fixed more nitrogen than the corresponding cultures of crop three.

Cation and Phosphorus Intake—The calcium contents, as are given in Table 21, were very low, both in percentage and total. The concentrations were all below 0.45 per cent, the value suggested by Graham (11) as minimal for nitrogen fixation by soybeans in similar experiments. Magnesium was also very low, both in percentage and total, but was above the level supplied by the seed. In every case, even though the plants lost some of their initial potassium, phosphorus, or nitrogen, yet the magnesium and the calcium were always present in amounts in excess of that delivered by the seed. These two nutrient elements were always taken from the soil and were not contributed to the soil by the seed. Phosphorus totals showed a net loss to the soil in all cultures.

Carbohydrate Production—It is of interest and of considerable importance to notice the composition of the dry matter of plants which did not fix nitrogen in significant amounts, because of the extremely low level of bases and of phosphorus. The deficient supply of bases and phosphorus did not cause a cessation of photosynthetic activity since the reducing sugar contents as given by Table 22, were of the same order of magnitude as those of the preceding crop. The failure of the plants to fix nitrogen in quantity had not, at the time of harvest, been reflected in a lower content of reducing and total sugars, as was the case in crop one—Series B which was not nodulated.

The percentages and totals of starch were surprising. After two croppings, the cultures initially treated with increments of potassium showed increasing starch contents both in percentage concentration and in total. This effect was brought about, not by a greater movement of potassium from the soil into the plant, but, by a decrease in the amount lost from the plant back to the soil. Since nitrogen was not being fixed, and an inconsiderable quantity of sugar was needed for protein synthesis, consequently the increase in carbohydrates produced as a result of greater amounts of potassium metabolized was stored as starch. Cultures 1, 2, 3, and 4 contained 7.16, 8.38, 11.21, and 12.84 per cent respectively of this particular carbohydrate.

TABLE 21: Composition of plant materials as influenced by declining fertility. (Data for 50 plants. Crop three—series B.)

Culture No.	Application of Potassium M E	Calcium		Magnesium		Phosphorus		Barium	
		Per Cent	Total mgms.	Per Cent	Total mgms.	Per Cent	Total mgms.	Per Cent	Total mgms.
1	0	0.39	46.0	0.156	18.6	0.232	27.7	0.72	86.2
2	5	0.41	50.7	0.183	22.7	0.236	29.2	0.42	51.1
3	10	0.27	39.5	0.144	20.9	0.204	29.7	0.10	15.3
4	15	0.30	37.3	0.162	20.0	0.249	30.4	...	...

Fifty soybean seeds contained 181 mgms. of potassium. 8.9 mgms. of calcium. 11.5 mgms. of magnesium. 32.8 mgms. of phosphorus and no barium.

TABLE 22: Carbohydrate contents of plant materials as influenced by declining fertility. (Data for 50 plants. Crop three—series B.)									
Culture No.	Sugar				Starch		Hemicellulose		% Reducing Sugar of Total
	Reducing		Total						
	Per Cent	Total mgms.	Per Cent	Total mgms.	Per Cent	Total mgms.	Per Cent	Total mgms.	
1	6.38	761	9.08	1081	7.16	855	10.52	1258	70.3
2	5.55	683	8.08	995	8.38	1039	9.90	1225	68.7
3	6.06	874	8.20	1191	11.21	1625	10.46	1518	73.8
4	4.66	569	6.86	844	12.84	1801	10.24	1270	67.9

Significance of Further Fertility Decline as Revealed by Plant Performance—These starch contents were much higher than those of the nitrogen fixing crops, but were of about the same magnitude as those of the non-nodulated crops. Both crops probably stored starch as the result of a lack of nitrogen, but the fundamental causes for their nitrogen hunger were widely different. In one, all the fertility elements were present in abundance, but active legume bacteria were withheld. In the other, legume bacteria were present, but the fertility elements, particularly calcium, magnesium, and phosphorus were too low to permit symbiotic fixation. However, this important fact stands out, namely, that under the level of bases and phosphorus, which was so low that nitrogen fixation did not occur in appreciable quantities, yet the plants were still able to carry on photosynthesis and to produce appreciable yields of dry matter that were much higher in carbohydrate relative to protein than is the case in the normal nitrogen-fixing plants. A comparison of the carbohydrate carbon to nitrogen ratios of the different crops is shown graphically in Fig. 8.

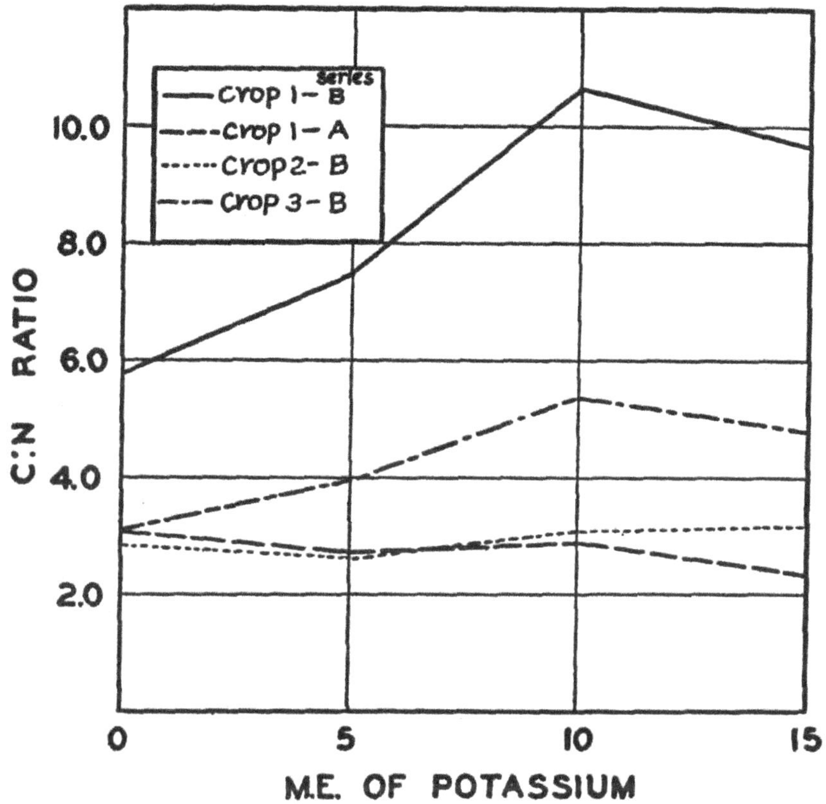

Fig. 8. Ratios of available carbohydrate carbon to nitrogen of plant material as influenced by variable potassium and different levels of fertility.

Under declining soil fertility, the plant's protein manufacturing processes are evidently the first to discontinue. Carbohydrate production, even in legumes, apparently continues under a much lower supply of fertility elements since the plants, which showed net losses of phosphorus, and of potassium, and which had very low supplies of nitrogen, magnesium, and calcium, were still able to produce appreciable quantities of starches and sugars.

Soil Exhaustion

The efficiency of a nutrient cation, when applied to the soil, is closely related to the rate of its removal and degree to which it is exhausted by the feeding activities of plants. Soil factors which have been shown to influence the rate and extent of fertility removal by crops, include among others: (1) the total amount of cation present in an available form; (2) the degree of saturation of the soil by the cation; and (3) the amount and nature of the other cations present. (16). The Series B was used to study the effect of continuous cropping in relation to the above factors. Changes in the soil were easily followed, since records were kept of all nutrient additions to the soil by titration and by the seed, and of nutrient removals in the crops. The difference between the amount of a nutrient element harvested and that supplied was considered to be the amount left in the culture medium in the available form. This assumption is not entirely correct since small quantities of nutrients may pass from the unavailable form in the clay crystal into an available form. However, the extent of this change (3) is so small as to preclude this possibility as a source of significant supplies of plant nutrients during a short period of five weeks. It is also possible that nutrients in an available or exchangeable form may become unavailable.

Depletion of Bases in the Soil—Continuous cropping reduced the level of bases, other than barium, in cultures 1, 2, 3, and 4 from saturations of 50, 66.6, 83.3, and 100 per cent respectively to very low levels. All cultures, after the third crop, were saturated to but 25 to 35 per cent with bases other than barium as is shown by Fig. 9. This low level was approached regardless of the initial treatment. Cultures supplied with higher amounts of potassium declined most rapidly in base saturation since potassium was most easily taken up. The critically low level of bases present in the soil was not without its effect on crop performance, as was indicated. As further evidences of the extreme unsaturation of the media, there are the very high degrees of acidity in all cultures, as given by the pH values in Table 23. All cultures had pH values below 4.5 after the third crop. Base saturation changes of each culture as influenced by cropping are shown graphically by Fig. 10 in which the outstanding feature is the increase in saturation by hydrogen with successive cropping.

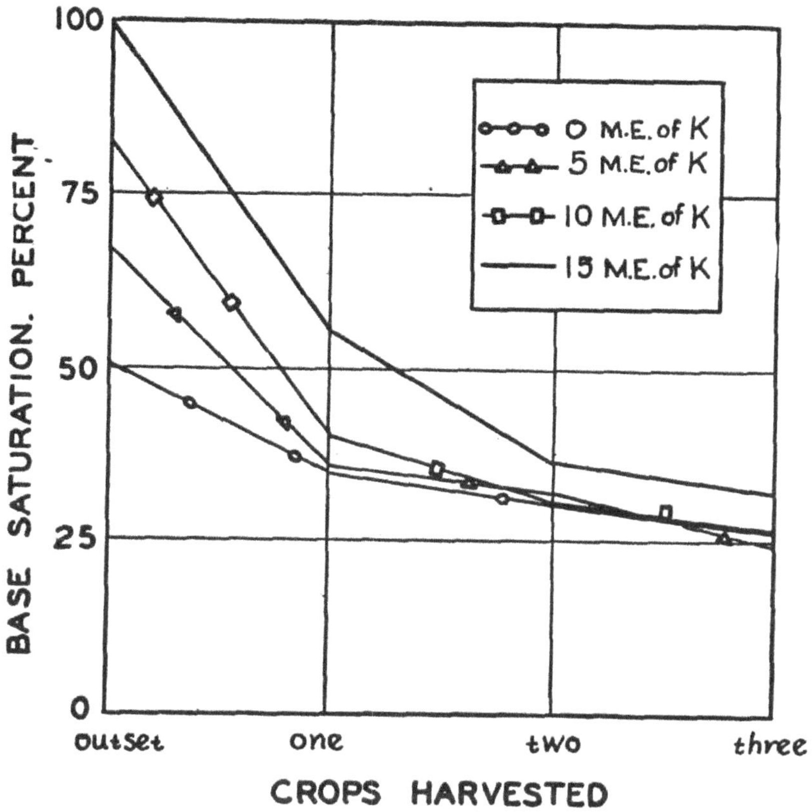

Fig. 9. Decline of bases, other than barium, as influenced by continuous cropping.

TABLE 23: Different degrees of acidity, pH, as a result of successive crops.					
Culture No.	Application of Potassium M E	Outset	After First Crop	After Second Crop	After Third Crop
1	0	6.60	5.50	5.02	4.44
2	5	6.70	5.34	4.80	4.25
3	10	6.75	5.05	4.55	4.25
4	15	6.75	4.50	4.40	4.20

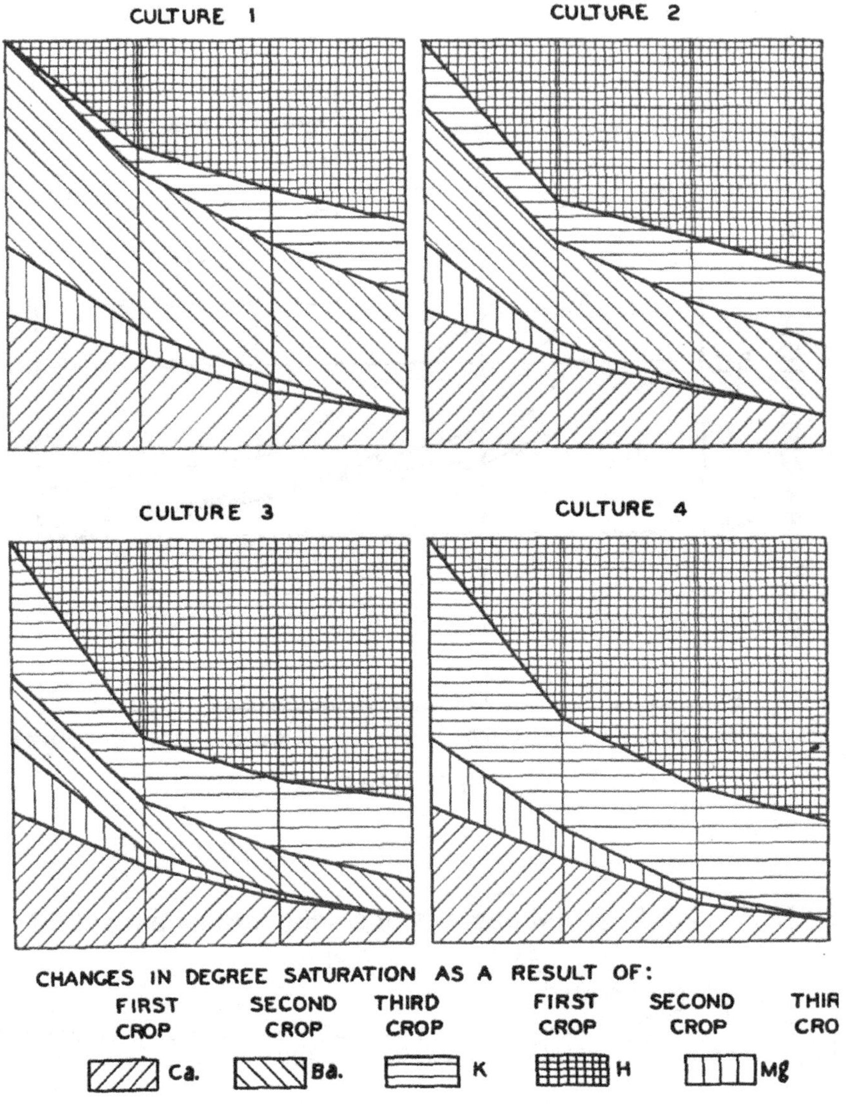

Fig. 10. Changes in the degree of saturation by the different cations as a result of successive croppings.

Potassium Exhaustion—In no case was there any potassium removed from the soil to which none had been added by titration. In fact, potassium moved in the reverse direction and the soil of culture 1 gained potassium continually from growing the three crops until it contained about 5.23 M E or an amount equal to 17.3 per cent of the exchange capacity. Culture 2 showed a decline in potassium saturation after the first crop, but continued cropping resulted in an increase in soil potassium. Regardless of the degree of initial potassium saturation, there was a marked tendency for all cultures to attain approximately the same value after continuous cropping.

TABLE 24: Potassium removal by three successive crops.

Culture No.	Crop 1			Crop 2			Crop 3			Potassium Remaining mgms.	Per Cent Recovery of Potassium
	Total K Supplied* mgms.	Taken mgms.	Per Cent Taken	Total K-Supplied** mgms.	Taken mgms.	Per Cent Taken	Total K Supplied mgms.	Taken mgms.	Per Cent Taken		
1	178	111.2	62.3	247.8	91.2	36.7	337.6	133.8	39.8	204	...
2	373.7	258.5	69.2	296.2	105.8	35.7	371.4	160.6	43.3	211	–7.8
3	569.4	379.4	66.7	370.8	157.0	42.3	394.8	151.8	38.4	243	62.2
4	765.1	456.3	58.9	495.4	195.9	39.6	480.5	193.8	40.4	287	48.9

*Includes titrated potassium plus the seed supply of 178 mgms.
**Titrated potassium unused by Crop 1 plus seed potassium.

The recoveries, or the extents of exhaustion, of the titrated potassium of cultures 2, 3, and 4, initially supplied with 5,10, and 15 M E of Potassium were –7.8, 62.2, and 48.9 per cent respectively, as is shown by Table 24. It is contrary to anticipation that the per cent exhaustion of an applied element should be lowest in the cultures receiving the smallest supply. Potassium loss from the clay to the crop was little influenced by the total cation saturation, as is shown by Fig. 10. Potassium moved in that direction, first in the culture saturated with bases at the outset to 100 per cent.

The nature of the residual potassium in the clay was studied by determining the amounts that could be removed by exchange with neutral normal ammonium acetate. A significant quantity was in a non-exchangeable form since only 19.0, 27.5, 24.1, and 25.8 per cent of the remnant of that originally supplied plus that added by crop growth, in the cultures 1, 2, 3, and 4 respectively were exchanged. These data are assembled in Table 25.

TABLE 25: Exchangeable potassium in the soil after three successive croppings.				
Culture No.	Application of Potassium	Remaining in Culture*	Exchangeable	
	M E	mgms.	mgms.	Per Cent
1	0	204	38.7	19.0
2	5	211	58.2	27.5
3	10	243	58.5	24.1
4	15	287	74.2	25.8

*Includes remnant of potassium supplied by the initial titration and that contributed by the seeds of crops grown.

Nitrogen Changes—The analyses for total nitrogen of the electrodialyzed clay soils before and after cropping show significant increases in nitrogen content as a result of cropping as given in Table 26. All cultures treated with potassium showed greater increments of nitrogen in the media than was shown by the culture with no treatment. Whether nitrogen was lost from the nodulated crops cannot be said since all cultures of the non-nodulated crop lost more nitrogen than was shown as final gain in nitrogen by any soil. These differences may have been due to volatilization of nitrogen from the media. Regardless of the crop or crops from which it came, the significant item is that nitrogen must have passed from the plant to the clay in a soluble state since there was no nitrogen contributed by aerial plant parts, and the amount of material added by sloughing off of root tissue was probably negligible.

TABLE 26: Changes in the total nitrogen in the soil after successive cropping. (Media of Series B.)				
Culture No.	Application of Potassium	Nitrogen		Increase in Nitrogen mgms.
	M E	At Outset mgms.	After Cropping mgms.	
1	0	68	102	34
2	5	68	115	47
3	10	68	120	52
4	15	68	116	48

Phosphorus Exhaustion—The efficiency of phosphorus absorption into the plant was stimulated by potassium additions. In the three crops, with a total of twelve cultures, there was only one instance in which an increase in applied potassium failed to give a greater percentage of supplied phosphorus taken. The data are given in Table 27. The third crop, which was growing under a very low level of bases and in media of extremely low pH, still showed consistent increases in efficiency of phosphorus removal with increased potassium. Cultures 1, 2, 3, and 4 took up 40.0, 48.3, 36.0 and 63.8 per cent respectively of the total phosphorus supplied by the seed and by the residual titrated phosphorus. Although the percentage taken of the total supplied showed consistent increases in all crops with potassium treatment, yet actually cultures 1 and 2 of crop two and all cultures of crop three lost phosphorus back to the soil as is shown by the data of Table 27 and graphically in Fig. 11. The percentage recovery of the initially supplied phosphorus increased tremendously with potassium treatment. In cultures 1, 2, 3, and 4 the amounts as percentage phosphorus recovery were –9.74, 17.6, 38.7, and 54.8 per cent respectively.

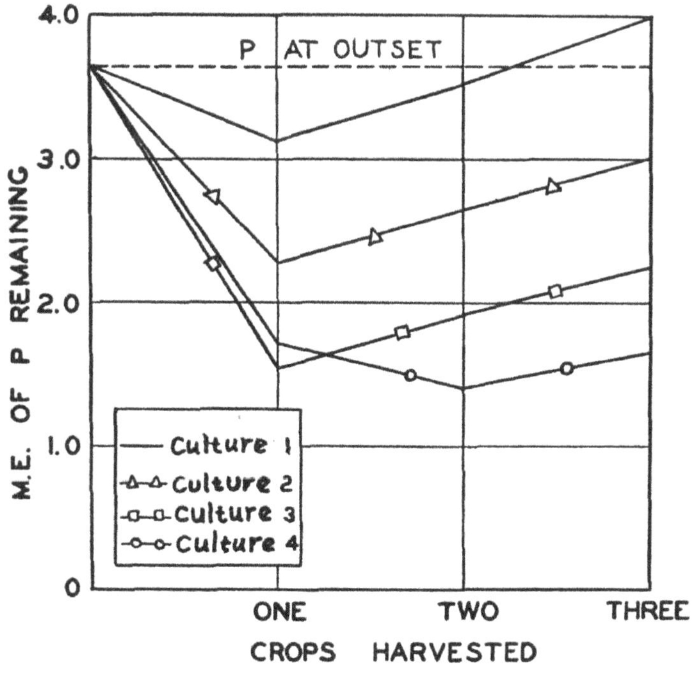

Fig. 11. Losses and gains in phosphorus by clay, resulting from successive cropping.

TABLE 27: Phosphorus removal by successive cropping. (Series B.)

Culture No.	Crop 1			Crop 2			Crop 3			Phosphorus Remaining in Cultures mgms.	Per Cent Recovery
	Total Sup-plied* mgms.	Taken mgms.	Per Cent Taken	Total** Supplied mgms.	Taken mgms.	Per Cent Taken	Total Supplied mgms.	Taken mgms.	Per Cent Taken		
1	61.3	29.3	47.4	65.3	28.5	43.7	69.4	27.7	40.0	41.7	-9.74
2	61.8	38.2	61.8	56.4	28.7	51.0	60.5	29.2	48.3	31.3	17.6
3	61.8	45.7	73.8	48.9	33.2	68.0	53.0	29.7	56.0	23.3	38.7
4	61.8	44.1	71.3	50.5	35.7	70.7	47.6	30.4	63.8	17.2	54.8

*The initial titrated supply was 38.0 mgms.
**Titrated phosphorus unused by preceding crop and seed phosphorus.

TABLE 28: Calcium removal by continuous cropping. (Series B.)

Culture No.	Crop 1			Crop 2			Crop 3			Calcium Remaining in Cultures mgms.	Per Cent Recovery
	Total Ca Supplied* mgms.	Taken mgms.	Per Cent Taken	Total Ca Supplied** mgms.	Taken mgms.	Per Cent Taken	Total Ca Supplied mgms.	Taken mgms.	Per Cent Taken		
1	208.5	66.7	31.9	150.7	67.1	44.5	92.5	46.0	49.8	46.5	76.7
2	208.5	80.4	38.5	137.0	56.0	40.8	89.9	50.7	56.4	39.2	30.5
3	208.5	92.2	44.2	125.2	60.2	48.0	73.9	39.5	53.5	34.4	82.8
4	208.3	88.7	42.5	128.7	78.2	60.8	59.4	37.3	62.6	22.1	39.0

* The initially titrated supply of Ca was 200.4 mgms. The seed content of Ca was 8.1 mgms.
**Titrated calcium unused by Crop 1 and seed calcium.

Calcium Exhaustion—Calcium recovery from the soil by the crop increased with potassium additions. The percentage recovery of the initial supply was rather high in all cultures as is indicated by the data of Table 28. Even in the cultures of lowest fertility there was no loss of calcium back to the soil.

Magnesium Exhaustion—Magnesium removal by continuous cropping with soybeans shows some very interesting results. From a constant initial supply of 60.8 mgms., cultures 1, 2, 3, and 4 recovered 95.1, 105, 107, and 98.4 per cent respectively. These data are given in Table 29 and shown graphically by Fig. 10. The three crops on cultures 2 and 3 took up 3.5 and 4.3 mgms. more of magnesium than the 60.8 mgms. originally supplied. These differences are probably within the limit of experimental error. Previous work (3), however, has suggested the possibility of a small magnesium release from the clay to the plant. It is also of interest that the soil at no time gained magnesium by loss from the planted seed. The extent and degree of magnesium exhaustion were very high in view of the fact that the base saturation of the soil was extremely low and showed pH values ranging from 4.5 to 4.2.

Barium Exhaustion—Barium was removed by cropping to a lesser extent than were the other divalent cations. Since it is a non-nutrient, one might expect it to behave in plant nutrition as though it were governed wholly by physico-chemical laws. If this is true, then absorption by the plant should be proportional to the concentration of barium present. That such was the case is strongly indicated by the data of Table 30, since the percentage recovery was almost a constant for all cultures, despite the fact that barium was supplied to cultures 1, 2, 3, in amounts varying as widely as 1:2:3. The percentage taken by each crop was approximately constant and the crop contained larger amounts as larger amounts were present. The efficiency of absorption by a given culture decreased considerably with the declining fertility level. Crops 1, 2, and 3 removed on an average 25.0, 14.5, and 10.4 per cent of the supplied amounts respectively.

TABLE 29: Magnesium removal by successive croppings. (Series B.)

Culture No.	Crop 1 Total Supplied* mgms.	Crop 1 Taken mgms.	Crop 1 Per Cent Taken	Crop 2 Total Supplied** mgms.	Crop 2 Taken mgms.	Crop 2 Per Cent Taken	Crop 3 Total Supplied mgms.	Crop 3 Taken mgms.	Crop 3 Per Cent Taken	Magnesium Remaining in Cultures mgms.	Per Cent Recovery
1	71.6	51.8	72.3	31.3	21.2	67.8	21.6	18.6	86.1	3.0	95.1
2	71.6	55.6	77.7	27.5	19.8	72.0	19.2	22.7	118.0	-3.5	105.5
3	71.6	56.3	78.7	26.8	21.7	81.0	16.6	20.9	126.0	-4.3	107.0
4	71 6	43.6	60.8	39.5	30.0	76.0	21.0	20.0	95.2	1.0	98.4

* The initially titrated supply of Mg was 60.8 mgms. The seed content of Mg was 10.8 mgms.
**Titrated magnesium unused by Crop I and seed magnesium.

TABLE 30: Barium removal by successive croppings. (Series B.)

Culture No.	Crop 1 Total Supplied* mgms.	Crop 1 Taken mgms.	Crop 1 Per Cent Taken	Crop 2 Total Supplied mgms.	Crop 2 Taken mgms.	Crop 2 Per Cent Taken	Crop 3 Total Supplied mgms.	Crop 3 Taken mgms.	Crop 3 Per Cent Taken	Barium Remaining in Cultures mgms.	Per Cent Recovery
1	1030.2	216.3	21.0	813.9	114.8	14.1	699.1	86.2	12.3	612.9	40.5
2	686.8	181.4	26.4	505.4	73.7	14.6	431.7	51.1	11.8	380.6	44.6
3	343.4	94.5	27.6	248.9	36.9	14.8	212.0	15.3	7.22	196.7	42.8
4	...	...	...	...	...	...	...	...	...	...	...
	Average—		25.0			14.5	...		10.4		

* The seed contained no barium.

Iron, Aluminum, and Silicon Removal by Plants—The intake and exhaustion of the non-nutrient cation, barium, from the adsorbed state suggested that perhaps silicon, aluminum, and iron behaved similarly. Considerable quantities of these elements were present in the electrodialyzed colloidal clay as shown by Table 1. Although the clay had been electrodialyzed, the data of Table 31 show that significant quantities of these elements were taken in by the plants. The content of iron decreased both in percentage and in total in all crops with increasing potassium supply. The silicon concentration in crop one showed a downward trend with treatment, but in the second and third crops silicon showed no definite trends. It is of interest that the total silicon, iron, and aluminum removed declined with successive croppings, as is shown by Table 32. This decrease cannot be ascribed to a decline in the total silicon, aluminum, and iron in the media, since the amounts removed by cropping represent only a very small fraction of the total amounts present. Neither is it due entirely to diminished yields of harvested materials for the decline in the removal of these elements is proportionally greater than the decrease in yields. The third crop produced as much dry matter as the second crop yet contained less of silicon, aluminum, and iron.

TABLE 31: Silicon, aluminum, and iron contents of plant materials as influenced by potassium. (Data for 50 plants. Series B.)

Culture No.	Application of Potassium M E	Silicon		Aluminum		Iron	
		Per Cent	Total mgms.	Per Cent	Total mgms.	Per Cent	Total mgms.
			First Crop				
1	0	0.42	64.1	0.19	27.2	0.420	60.1
2	5	0.33	64.4	0.20	39.0	0.230	45.9
3	10	0.26	58.1	0.14	32.6	0.075	16.9
4	15	0.27	56.0	0.16	33.3	0.086	18.2
			Second Crop				
1	0	0.18	19.0	0.12	13.1	0.29	31.0
2	5	0.19	19.6	0.09	9.9	0.19	20.4
3	10	0.17	19.4	0.07	7.9	0.24	27.8
4	15	0.21	29.1	0.12	17.0	0.14	19.3
			Third Crop				
1	0	0.18	22.0	.064	7.8	0.115	13.8
2	5	0.18	22.2	.074	9.2	0.098	12.1
3	10	0.16	23.3	.087	12.6	0.080	11.6
4	15	0.13	15.1	.066	6.8	0.052	5.6

Fifty seeds contained 0.2 mgms. of iron and only traces of silicon and aluminum.

TABLE 32: Removal of silicon, aluminum, and iron as influenced by successive croppings.			
Crop	Silicon mgms.	Aluminum mgms.	Iron mgms.
First	242.6	131.1	141.1
Second	87.1	47.9	98.5
Third	82.6	36.4	43.1

There is suggested the possibility that a part of the silicon, aluminum, and iron of the electrodialyzed clay is present in a form more readily taken in by plants than the major portion. If such were the case, then these elements too would be subject to depletion by successive croppings. This suggestion is supported by some work by Mattson (28) who showed that a part of the iron and aluminum even in an electrodialyzed clay can be liberated simply by washing the clay with solutions of neutral salts. Successive treatments with salt solutions removed correspondingly less of aluminum and iron.

Significance of Soil Depletion as Reflected by Plant Performances—That fertility decline under cropping proceeds at a decreasing rate until a fairly constant level is obtained is a common observation. However, the reverse process, or the building up to that level, by cropping is not commonly observed. The results obtained in the study of the soil exhaustion indicate that this was happening in cultures 1 and 2, with respect to potassium. The ready passage of potassium from living plant roots to an unsaturated colloidal clay has been demonstrated by Jenny and Overstreet (20), working with barley roots. They showed that the rate of depletion was much slower with a saturated clay. The data obtained with soybeans in this study indicate that this reverse movement of potassium, from plants high in this constituent, is not dependent entirely on a highly unsaturated medium, since the reversal of potassium movement occurred from plants grown five weeks, even in soil saturated to 100 per cent initially with barium, magnesium, and calcium. One would infer that in a highly adsorptive medium, such as the soil, a minimal level of available potassium is necessary to prevent the loss of potassium from plants high in this element. The extent of recovery of the initially supplied elements shows some very interesting differences as shown by the data assembled in Table 33. The divalent ions moved into the plants in all cases even though the total amount and per cent saturation of the ion on the clay were very low. This was particularly true in the media of the last crop. Calcium behavior was similar to that of magnesium. Both showed a net movement into the plant. Barium was absorbed to a lesser degree than either magnesium or calcium.

TABLE 33: Initially supplied nutrients recovered by successive croppings.

Culture No.	Application of Potassium M E	Potassium %	Phosphorus %	Calcium %	Magnesium %	Barium %
1	0	...	-9.74	76.7	95.1	40.5
2	5	-7.8	17.6	80.5	105.5	44.6
3	10	62.2	38.7	82.8	107.0	42.8
4	15	48.9	54.8	89.0	98.4	...

The monovalent ion, potassium, showed no consistency in direction of movement, going into the plant when sufficient quantities in an available form were supplied in the soil and being lost to the soil under other conditions. Phosphorus and potassium behaved similarly as regards gain or loss by the plant. The total phosphorus removed by cropping was stimulated considerably by potassium treatment.

Nitrogen gains in the cultures were due, in part at least, to losses from crop one, the non-nodulated crop. At the time of harvest this crop was showing marked symptoms of nitrogen deficiency. It seems highly probable that protein nitrogen was being broken into cationic nitrogen, a form which would have a great tendency to unite with the colloidal substrate. The passage of nitrogen from living plant nodules into the substrate has been reported by Virtanen (40) (41). He reported this movement to be greatest with plants growing on substrates such as soil or kaolin.

It is suggested by the data that magnesium intake can occur at levels of saturation so low that phosphorus and potassium are being lost from the plant. The data also suggest that the percentage intake of total magnesium supplied does not decrease greatly under declining fertility levels as was true for potassium. The fact that no loss of seed magnesium occurred under extremely low levels of magnesium saturation emphasizes the properties by which magnesium, once metabolized, must remain within the plant tissue.

The proportionality between barium absorption and the amounts supplied suggests that there was no selective influence by the plant affecting barium intake. Such a suggestion might also be drawn from the fact that barium is not considered as playing any functional role in plant metabolism. Within limits, barium exerted no recognizable injurious effect on plants grown in substrates high in colloid content. From these observations, the use of barium as a cation to be varied reciprocally with any other cation seems to be a justifiable experimental technique.

Discussion

The results obtained in this study show a definite influence by potassium on nitrogen fixation and plant composition. The total nitrogen content of nodulated soybeans which were well supplied with calcium, magnesium, and phosphorus, increased with increasing potassium treatment. However, its percentage concentration remained almost a constant or decreased slightly with increasing potassium supplied indicating a greater effect of potassium on carbohydrate production. The increased absorption of calcium and phosphorus in the potassium treated cultures prevented these elements from being limiting factors in nitrogen fixation.

From the data presented there is evidence that potassium is a factor in nitrogen fixation and protein formation not only through its stimulation of carbohydrate production but, since the reducing sugars declined while other carbohydrates, such as starch and hemicelluloses, remained constant with increasing potassium additions, it might be inferred that potassium played an additional role in the protein synthesis.

Plants which were inoculated and fixing nitrogen showed no significant changes in the ratio of carbohydrate to protein in the plants at the higher levels of potassium used in the experiment, since proportionately more nitrogen was fixed. However, when nitrogen was deficient either because of non-nodulation or of an insufficient supply and degree of saturation of the other fertility elements, potassium treatments resulted in a tremendous increase in carbohydrate production and storage so that the ratio of carbohydrate to protein widened considerably as a consequence. This is of considerable significance in the production and use of legumes for purposes of forage feed.

All soybeans which were non-nodulated had lost an average of 23 per cent of their original seed nitrogen at the end of a growing period of five weeks. This loss was probably accelerated by the very high absorptive capacity of the media for soluble cationic substances. Soluble cationic nitrogen is found in plants in more or less amounts during the entire growing period. The fact that these plants had just begun to show nitrogen deficiency symptoms may also have been a factor in nitrogen loss since it has been shown (12) that during senescence of plants there is a breakdown of plant proteins into less complex mobile substances some of which presumably might possess cationic properties. Nitrogen losses from nodules of cowpeas to the media have been extensively studied by Virtanen and co-workers (40) (41). These excreted nitrogen compounds were shown to consist largely of amino acids. He found no evidence of nitrogen excretion from root surfaces of non-nodulated cowpeas which were grown in sand or kaolin.

Since calcium and magnesium are chemically related they might be expected to behave similarly in the processes of nutrient absorption and metabolism by the plant. That there is a similarity in certain performances was indicated, since the data show that in all cases, even under very low levels of fertility, calcium and magnesium always showed a net gain in the plants above the seed contents. There is evidence that the feeding capacities of soybean plants for magnesium and calcium, when supplied in an exchangeable form on a colloidal clay, differ considerably. The magnesium taken as percentage of the total supplied, was, in all cases from one and one-half to two times as great as that for calcium. Magnesium absorption remained high when the magnesium supply and the supply of other bases was so low that phosphorus and potassium were lost from the plants to the soil. The relative positions of calcium and magnesium in the ionic double layer of the exchange complex as it affects their ease of replacement may serve to explain these differences. Magnesium has been shown to be more highly hydrated than calcium and more readily displaced from homionic clays of calcium and magnesium (10). This explains the common observation that plants may grow on soils with an apparently low content and degree of saturation of exchangeable magnesium and yet show no significant response to applications of magnesium.

Summary and Conclusions

A study was made of the influence of potassium on nitrogen fixation and plant composition of soybeans. An electrodialyzed colloidal clay from the Putnam subsoil was used as a carrier of this cation along with constant amounts of calcium, magnesium, phosphorus, and non-nutrient barium varying reciprocally with potassium.

The effect of fertility depletion by continuous cropping on nitrogen fixation and carbohydrate production was also studied. Attention was given to the rate and degree of depletion from the soil of the various elements by successive croppings of soybeans. On the basis of this work the following facts seem of importance.

1. Nitrogen fixation increased with higher levels of available potassium when accompanied by constant amounts of calcium, phosphorus, and magnesium.
2. The concentration of reducing sugars in both nodulated and non-nodulated crops decreased with increasing potassium treatment.
3. In non-nodulated soybeans the starch content increased tremendously, both in percentage and totals, with the higher levels of available potassium. Soybeans which were nodulated and fixing nitrogen increased in total starch but remained relatively constant in percentage concentration as increments of potassium were supplied.
4. Crops on media which were too low in the fertility items to permit nitrogen fixation and protein production were almost constant in total nitrogen content but contained increasing amounts of carbohydrates, especially starch, in proportion to the amount of potassium metabolized.
5. Soybeans lost potassium from the seed to the soil in cultures with no initial treatment of potassium, even though the soil was fully saturated with bases other than potassium and approximately neutral at the outset.
6. Potassium was taken by plants in relatively large quantities from soil initially saturated with bases and containing potassium in an exchangeable form.
7. The efficiency of phosphate removal by plants was enhanced considerably by potassium treatments.
8. The plants showed a greater relative absorption and depletion of magnesium than of any other cation.
9. All media tended to approach the same degree of saturation of ions other than barium when cropped successively.
10. Potassium, phosphorus, and nitrogen moved from the plant to the soil under certain conditions. The cations, calcium, and magnesium were always present in the harvested crop in excess of the amounts supplied by the seed.

11. Significant quantities of nitrogen were lost back to the soil by non-nodulated soybeans even in the relatively short period of five weeks.
12. Crops growing under a low supply of nitrogen and an adequate supply of bases produced more root growth in relation to tops than crops which were nodulated and which fixed nitrogen.

References

1. Albrecht, W. A. *Calcium and Hydrogen Ion Concentration in the Growth and Inoculation of Soybeans*. Jour. of the Amer. Soc. of Agron. 24: 793–806. 1932.
2. Albrecht, W. A. and McCalla, T. M. *The Colloidal Clay Fraction of Soil as a Cultural Medium*. Amer. Jour. Bot., 25: 403–407. 1938.
3. Albrecht, W. A., Graham, E. R., and Ferguson, C. E. *Plant Growth and the Breakdown of Inorganic Soil Colloids*. Soil Science, 47: 455–458. 1938.
4. Alten, F. and Goeze, G., *Versuche uber die Wirkung des Kaliums auf die Kolensaureaufnahme von Wiesenblattern*. Die Ernahrung der Pflanze, 33: 21–28. 1937.
5. Bradfield, R. *The Chemical Nature of a Colloidal Clay*. Mo. Agr. Exp. Sta. Res. Bul., 60. 1923.
6. Briggs, G. E. *Experimental Researches on Vegetable Assimilation and Respiration. XVI. The Characteristics of Subnormal Photosynthetic Activity Resulting from Deficiency of Nutrient Salts*. Royal Soc. of London, B: 94: 20–34. 1922–23.
7. Drake, M. and Scarseth, G. D. *Relative Abilities of Different Plants to Absorb Potassium and the Effects of Different Levels of Potassium on the Absorption of Calcium and Magnesium*. Proc. of the Soil Soc. of America, 4: 201–204. 1939.
8. Fonder, J. F. *Variations in Potassium Content of Alfalfa Due to Stage of Growth and Soil Type and the Relationship of Potassium and Calcium, in Plants Grown upon Different Soil Types*. Jour. of the Amer. Soc. of Agron., 21: 732–750. 1929.
9. Gieseking, J. E., Snider, H. J., and Getz, C. A. *Destruction of Organic Matter in Plant Material by the Use of Nitric and Perchloric Acids*. Ind. and Eng. Chem., Anal. Ed., 7: 185. 1935.
10. Gieseking, J. E., and Jenny, H. *Behavior of Polyvalent Cations in Base Exchange*. Soil Science, 42: 273–280. 1936.

11. Graham, E. R. *Magnesium as a Factor in Nitrogen Fixation by Soybeans*. Mo. Agr. Exp. Sta. Bul., 288. 1938.
12. Gregory, F .D. *Mineral Nutrition of Plants*. Ann. Rev. of Biochem., 6: 557. 1937.
13. Hartt, C. E. *Some Effects of Potassium Upon the Amounts of Protein, Sugars and Enzyme Activity of Sugar Cane*. Plant Physiol., 9: 452–490. 1934.
14. Heinze, P. H., and Murneek, A. E. *Comparative Accuracy and Efficiency in Determination of Carbohydrates in Plant Material*. Mo. Agr. Exp. Sta. Res. Bul., 314. 1940.
15. Hoagland, D. R., and Martin, J. C. *Absorption of Potassium by Plants in Relation to Replaceable, Nonreplaceable and Soil Solution Potassium*. Soil Sci., 36: 1–35. 1923.
16. Horner, G. M. *Relation of the Degree of Base Saturation of Colloidal Clay by Calcium to the Growth, Nodulation, and Composition of Soybeans*. Mo. Agr. Exp. Sta. Res. Bul., 232. 1936.
17. Janssen, G., and Bartholomew, R. P. *The Translocation of Potassium in Tomato Plants and its Relation to their Carbohydrate and Nitrogen Distribution*. Jour. of Agr. Res., 38: 447–465. 1929.
18. Janssen, G., and Bartholomew, R. P. *The Influence of the Potash Concentration in the Cultural Medium on the Production of Carbohydrates in Plants*. Jour. of Agr. Res., 40: 243–261. 1930.
19. Janssen, G., and Bartholomew, R. P. *The Effect of Potassium on the Production of Proteins, Sugars and Starch in Cowpea and Sugar Beet Plants and the Relation of Potassium to Plant Growth*. Jour. Amer. Soc. of Agron., 24: 667–679. 1932.
20. Jenny, H., and Overstreet, R. *Cation Interchange Between Plant Roots and Soil Colloids*. Soil Sci., 47: 257–272. 1939.
21. Johnston, E. S., and Hoagland, D. R. *Minimum Potassium Level Required by Tomato Plants Grown in Water Cultures*. Soil Sci., 27: 89–109. 1929.
22. Kolthoff, T. M., and Sandell, E. B. *Textbook of Quantitative Inorganic Analysis*. Macmillan Company, New York. 1937.
23. Kraus, E. J., and Kraybill, R. R. *Vegetation and Reproduction with Special Reference to the Tomato*. Ore. Agr. Col. Exp. Sta. Bul., 149. 1918.
24. Loew, O. *Zur Physiologie von Kalium und Kalzium*. Die Ernahrung der Pflanze, 30: 141–144. 1934.
25. Lundegardh, H. *Die Nahrstoffaufnahme der Pflanze*. Jena. 1932.

26. Maiwald, G. *Methodisches zur Beteiligung des Kaliums an der Stoffzengung der Hoheren Pflanzen. Der Forschungsdienst,* 7:121–128. 1938.
27. Marshall, C. E. *Layer Lattice and Base Exchange Clays*. Ztschr. Krist, A: 91: 433–449. 1935.
28. Mattson, S. *Electrodialysis of the Colloidal Soil Material and the Exchangeable Bases*. Jour. of Agr. Res., 33: 553–567. 1926.
29. McCallum, A. B. *On the Distribution of Potassium in Animal and Vegetable Cells*. Jour. of Physiology, 32: 95–128. 1905.
30. Murneek, A. E., and Heinze, P. H. *Speed and Accuracy in Determination of Total Nitrogen*. Mo. Agr. Col. Exp. Sta. Bul., 261. 1937.
31. Nobbe, F., Schroder, J., and Erdmann, R., *Ueber die Organische Leistung des Kaliums in der Pflanze*. Landw. Versuchs. Stat., 13: 321–423. 1871.
32. *Official and Tentative Methods of Analysis of the A. O. A. C.* Washington, D.C. 1935.
33. Overstreet, R., and Jenny, H. *Studies Pertaining to the Cation Absorption Mechanism of Plants in Soils*. Proc. of the Soil Sci. Soc. of Amer. 4: 125–130. 1939.
34. Paech, K., *Uber die Regulation des Eiweissumsatzes und Uber den Zustand der Proteolytischen Fermente in den Pflanzen*. Planta, 24: 78–129. 1935.
35. Petrie, A. H. K. *The Drift of the Content of Potassium and Calcium with Age in Plants*. Australian Jour. of Exp. Biol. and Med. Sci., 12: 99–110. 1934. Chem. Abst., 29: 2575.
36. Rauterberg, E., and Loofmann, U. H. *Beziehungen Zwischen der Ernahrung und der Chemischen Zusammensetzung der Pflanzen*. Der Forschungsdienst 7: 203–210. 1938.
37. Reed, H. S. *The Value of Certain Nutritive Elements to the Plant Cell*. Ann. Bot., 21: 501–543. 1907.
38. Stocklasa, J. *Ist das Kaliumion an der Eiweiss-synthese in der Pflanzen Zelle Beteiligt?* Biochem. Ztschr., 73: 107–160. 1916.
39. Thomas W. H. *The Reciprocal Effects of Nitrogen, Phosphorus, and Potassium as Related to the Absorption of These Elements by Plants*. Soil Sci., 33:1–17. 1932.
40. Virtanen, A. I., Von Hausen, S., and Laine, T. *Investigations on the Root Nodule Bacteria of Leguminous Plants. XIX, Influence of*

Various Factors on the Excretion of Nitrogenous Compounds from the Nodules. Jour. of Agr. Sci., 27: 332–348. 1937.

41. Virtanen, A. I., Von Hausen, S., and Laine, T. *Investigations on the Root Nodule Bacteria of Leguminous Plants. XX, Excretion of Nitrogen in Associated Cultures of Legumes and Non-Legumes*. Jour. of Agr. Sci. 27: 584–610. 1937.
42. Wilcox, L. V. *Determination of Potassium Using an Aqueous Solution of Trisodium Cobaltinitrite*. Ind. and Eng. Chem., Anal. Ed., 9: 136. 1937.

CHAPTER 14

Nitrate Absorption by Plants as an Anion Exchange Phenomenon[1]

Recent studies indicating the importance of cation exchange from the clay of the soil to the root of the plant, in what has been suggested as contact exchange by Jenny and Overstreet (1938) and demonstrated by means of calcium and hydrogen systems by Albrecht (1940), have been exceedingly helpful to the agronomist in his interpretation of soil conditions and plant growth.

The agronomist has found the soil solution theory of plant nutrition, based upon nutrient solution studies, insufficiently applicable to soil conditions. This is not surprising when careful consideration is given to the physico-chemical state of a colloidal system such as occurs in most soils.

Work to date in plant nutrition by means of colloidal exchange has considered only the cation part of the plant's needs. The question now arises whether exchangeable anions, such as adsorbed nitrate, would be equally as available to plants as the adsorbed cations.

According to Hoagland and Arnon (1941) "The importance of the soil solution as a direct source of nutrients is manifested by the fact that an ion of major quantitative importance, namely, the nitrate ion, is present primarily in dissolved form and must be accompanied by equivalent amounts of cations." This is, of course, essentially true; yet the possibility still remains that there may be a source of adsorbed nitrate ions in the soil. In case soils should retain nitrate ions in the adsorbed form and such should be available for plant nutrition, this would be a very desirable condition. To date, techniques are not refined enough to separate adsorbed forms from soluble

[1] *E. R. Graham and Wm. A. Albrecht. "Nitrate Absorption by Plants as an Anion Exchange Phenomenon." American Journal of Botany 30, no. 3 (March 1943): 195–8.*

forms of nitrate. The adsorbed nitrate would not be leached from the soil profile, as is true for nitrate in solution. It was under the assumption that exchangeable anions might later be shown to be available in the soil for plant nutrition that the following investigation was undertaken.

Experimental Procedure

The following experiment was carried out in order to determine the anion exchange capacity of a commercial product known as Amberlite (IR4). A sample of this anion exchange material[2] was placed in a Gooch crucible, leached with a liter of distilled water and then leached with 500 ml. of normal nitric acid. This was followed by leaching with three liters of distilled water. The leachate was tested for absence of nitrates. One-half liter of normal sodium sulfate solution was then leached through the Amberlite and the leachate analyzed for nitrate. On the basis of the wet weight of Amberlite, the anion exchange capacity was found to be 160 M.E. of the nitrate anion per 100 grams of Amberlite. Calculations were also made for the oven dry basis, 100°C. The anion exchange capacity was 319 M.E. for nitrate anion per 100 grams of Amberlite, dried at this temperature.

Experiment A—A sample of Knox subsoil, which contained only .0277 per cent nitrogen, was obtained for the substrate. This soil was placed in stone jars of three gallon capacity and the moisture content and weights noted. In order to ensure that the plants would be adequately supplied with potassium and phosphorus, 10 M.E. of dipotassium phosphate were added to each jar. A series, number one, was made up, which had no other treatment than the above. Another series, number two, contained 15 grams of Amberlite per jar which had received a treatment of carbonic acid until saturated with carbonate ions. A third series contained, per jar, 15 grams of Amberlite which had been saturated with nitrate ions. Eight seeds of Missouri Hybrid WF 9 × 38–11, 1463 × 14542[3] maize were planted in each jar. After the plants were started they were reduced to four uniform plants in each jar. After a growing period of four weeks they were harvested, and the dry weights were noted.

2 Acknowledgment is gladly made to the Resinous Products and Chemical Co. of Philadelphia, Pa., for furnishing the Amberlite for this study.

3 The authors are indebted to L. Hogemeyer for furnishing the seeds used in this experiment.

Experiment B—A series of maize cultures was prepared in like manner. However, in this case, treatments of 8 and 16 M.E. of calcium nitrate in solution were added to some cultures, so that a comparison could be drawn between 8 M.E. of adsorbed nitrate and 8 M.E. of water soluble nitrate. These plants were grown for six weeks, harvested, dried, and the crop weights noted.

Experiment C—A series of three-gallon stoneware jars was arranged with a glass tube packed down the inside. The end of the tube was surrounded with glass wool so that the tube would not become clogged. This tube was designed to serve for water removal in the saturation and leaching of the soil. These jars were then carefully filled with Knox subsoil mixed with coarse sand. The soil in each jar was treated with 10 M.E. of dipotassium phosphate. Amberlite was added in sufficient quantity so that a series with 24 and 32 M.E. of adsorbed nitrate was supplied. To compare with this, a series of jars containing 24 and 32 M.E. of calcium nitrate in solution was prepared. A check series containing no added nitrogen was also set up. Maize kernels were planted and thinned as in Experiment A. After the plants had grown for a period of three weeks, the untreated cultures, the two cultures with Amberlite and the two cultures with calcium nitrate as solution were leached thoroughly and the leachate collected. Leaching was accomplished by adding an excess of water to the entire culture, then removing the water by means of a suction pump attached to the glass tube which reached into the bottom of the jar. The leachate was carefully collected and analyzed for nitrogen. After the plants had grown for a total period of six weeks—three weeks before and three weeks after the soils were leached—they were harvested, oven dried, dry weights noted and a standard Kjeldahl analysis used for the determination of the total nitrogen.

Results

Experiment A—The growth appearance of the plants in experiment A is shown in Plate 1 and the crop weights are recorded in Table 1. The plants grew poorly on the low nitrogen substrate and on the Amberlite saturated with carbonate. The plants on the Amberlite saturated with nitrate grew very well

Experiment B—The growth appearances of the crops of Experiment B are shown in Plate 2, and the crop weights are recorded in Table 2. The plants grown on cultures containing Amberlite saturated with nitrate grew as well as those on cultures containing water soluble nitrate nitrogen. In every case the plants treated with nitrogen grew much better than those with no treatment.

Plate 1. Maize plants as influenced by different exchange anions adsorbed on Amberlite. 1. No Amberlite. 2. Amberlite saturated with the carbonate anion. 3. Amberlite saturated with the nitrate anion.

TABLE 1: Weights of maize plants as influenced by different exchange anions adsorbed on Amberlite.

Culture number	Treatment	Total crop weight in gms.
1	None	1.060
2	15 grams Amberlite saturated with carbonate anion	1.160
3	15 grams Amberlite saturated with nitrate anion	19.315

Experiment C—The amounts of nitrogen which could be leached from cultures of maize plants growing for three weeks on the variously treated cultures are shown in Table 3. The data show that 14.1 M.E. of nitrate could be leached from a culture treated with 24 M.E. of calcium nitrate in solution, while only 1.1 M.E. could be leached from a culture containing 24 M.E. of adsorbed nitrate. Relatively similar results were obtained in the cultures treated with 32 M.E. of nitrogen.

Plate 2. Maize plants as influenced by water soluble or adsorbed nitrate nitrogen. 1. No treatment. 2. 8 M.E. of water soluble nitrogen. 3. 16 M.E. of water soluble nitrogen. 4. 8 M.E. of adsorbed nitrogen. 5. 16 M.E. of adsorbed nitrogen.

TABLE 2: Weights of maize plants as influenced by nitrogen as water soluble nitrate and as adsorbed nitrate.

Culture number	Nitrogen Added M.E.	Crop weight 4 plants
1	None	2.71 gms.
2	8 as calcium nitrate	9.70 gms.
3	16 as calcium nitrate	11.93 gms.
4	8 as (IR4) nitrate	10.41 gms.
5	16 as (IR4) nitrate	11.61 gms.

The nitrogen contents and the crop weights of the maize plants were changed greatly by leaching in the case of the plants grown on cultures treated with calcium nitrate in solution. On the other hand, leaching had little effect on either the nitrogen content or crop weights of plants grown on cultures treated with adsorbed nitrogen, as is evident from the data in Table 4 and shown graphically by Figure 1.

TABLE 3: Nitrogen leached from cultures growing maize plants for three weeks on substrates containing water soluble and adsorbed nitrogen.		
Culture number	Nitrogen Added M.E.	Nitrate nitrogen removed by leaching M.E.
384	None	.2
185	24 as calcium nitrate	14.1
202 A	32 as calcium nitrate	21.2
373	24 as (IR4) nitrate	1.1
156	32 as (IR4) nitrate	1.3

TABLE 4: Crop weights and nitrogen contents of the crop on leached cultures, supplied nitrogen in the water soluble and the adsorbed forms.				
Culture	Nitrogen Added M.E.	Crop weight /4 plants gms.	Nitrogen	
			Per cent	Crop total M.E.
284	None	2.335	.87	4.5
384	None (leached)	2.240	.89	4.3
334	24 as calcium nitrate	19.183	1.41	19.3
185	24 as calcium nitrate (leached)	10.480	.75	5.6
388	32 as calcium nitrate	18.925	1.78	24.1
202 A	32 as calcium nitrate (leached)	13.805	.93	9.1
345	24 as (IR4) nitrate	14.268	.94	9.6
373	24 as (IR4) nitrate (leached)	15.793	.90	10.2
3A	32 as (IR4) nitrate	21.750	1.03	16.1
156	32 as (IR4) nitrate (leached)	24.503	.96	16.8

Discussion

The results show definitely that adsorbed nitrate ions were available for plant nutrition. The adsorbed nitrogen in experiment A influenced the growth of maize so that the total crop weight was 19 times that of the maize growing on a subsoil that was low in nitrogen but which had been treated with potassium and phosphorus. Adsorbed nitrate nitrogen was as effective on plant growth as was the same quantity of nitrate as water soluble calcium nitrate added to the soil substrate. Leaching had little effect in retarding the growth of maize plants given adsorbed nitrogen, while it reduced the growth greatly and also the nitrogen content of the plants in the case of the cultures treated with water soluble nitrogen.

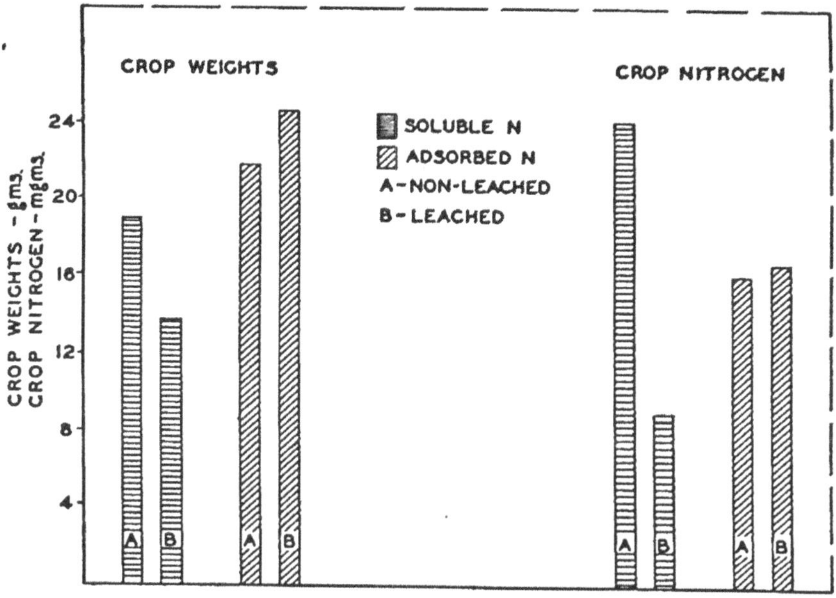

Fig. 1. Crop weights and crop contents of total nitrogen for maize offered soluble and adsorbed nitrogen in the substrate that was leached and unleached.

When a comparison is made of the nitrogen concentration in the plants grown on cultures containing water soluble nitrogen and in the plants grown on cultures containing adsorbed nitrogen, the differences are of interest (Table 5).

TABLE 5.

Culture number	Nitrogen Added M.E.	Nitrogen per cent	Crop weight 4 plants gms.	Nitrogen total mgms.
388	32 as calcium nitrate	1.78	18.925	33.6865
3a	32 as (IR4) nitrate	1.03	21.750	22.4025

It was not the purpose of this study to demonstrate the mechanism by which the adsorbed nitrogen was released from the Amberlite and made available for plant use. It appears, however, that one might propose the following hypothesis: According to Meyers, Eastes, and Meyers (1942) the synthetic resin, Amberlite, used in this study takes up nitric acid according to the following equation:

$$\{\text{Anion Exchange Resin Acid Absorbent}\} + HNO_3 \rightarrow \{\text{Anion Exchange Resin Acid Absorbent H}\}\ NO_3$$

In this case the hydrogen of the nitric acid is fixed and the nitrate ion left to be exchanged by any anion with which it might come in contact. Since plants during respiration release significant amounts of carbon dioxide to the soil solution, the carbonate anion could exchange for the nitrate anion according to the following equation:

$$\{\text{Anion Exchange Resin Acid Absorbent H}\}NO_3 + H_2CO_3 \rightarrow \{\text{Anion Exchange Resin Acid Absorbent H}\}CO_3 + 2HNO_3$$

The possibility arises that if very large amounts of carbon dioxide were produced by the plants at one instant, large amounts of nitrate anion would be rendered water soluble. The results of the leaching experiment showed that only small amounts of nitrate could be leached from the substrate even while the plants were actively growing. This indicates that at any given instant the carbonate anion was not replacing very large amounts of the nitrate.

The Amberlite used in this study contained 17 per cent of nitrogen (Meyers, et al. 1942). With this large amount of nitrogen contained in Amberlite, the question as to its availability for plant growth should be considered. Under the conditions of this experiment it appears that this nitrogen was not available to plants. The data show that when 15 grams of Amberlite, saturated with carbon dioxide, were added to the soil, the plants behaved as though they were grown on a substrate given no treatment.

Summary

A study of nitrate adsorption by plants as it was related to anion exchange phenomena revealed that:

1. Adsorbed nitrate ions were found to be available for plant nutrition.
2. Maize plants on a substrate of low nitrogen content grew equally well when extra nitrogen was added in the adsorbed form and when nitrogen was added in the water soluble form.

3. When plants were growing actively and water soluble nitrogen was leached from cultures given nitrogen in this form, such leaching resulted in a nitrogen deficiency and retarded growth by the crop. But when cultures containing adsorbed nitrogen were leached, no nitrogen deficiency in the crop could be induced, and no ill effect on the growth of maize plants resulted.

Literature Cited

1. Albrecht, Wm. A. 1940. Adsorption ions on the colloidal complex and plant nutrition. Soil Science Society of America 5: 8–16.
2. Hoagland, D. R., and D. T. Arnon. 1941. Physiological aspects of nutrients for plant growth. Soil Sci, 51: 431–444.
3. Jenny, H., and R. Overstreet. 1938. Contact effects between plant roots and soil colloids. Proc. Nat. Acad. Sci. U.S.A. 24: 384–492.
4. Meyers, Eastes and Meyers. 1942. Private communication from Mr. Frederick J. Meyers, Resinous Products and Chemical Co., Philadelphia, Pa.

CHAPTER 15

Nitrate Production in Soils as Influenced by Cropping and Soil Treatments[1]

Introduction

Though the soil may be judged in terms of its stock of plant nutrient elements, it must be considered also in terms of its microbiological activity by which these are converted into such forms as will serve the plants. Nitrogen is the nutrient element of foremost concern, both in terms of its stock in the soil and of its conversion from the promiscuous, less active, organic matter forms into the water-soluble, nitrate form; consequently, the influence of the long continued soil treatments on the conversion of this element becomes significant. The soil serves effectively as a factory to deliver a liberal supply of nitrate nitrogen to the growing crop regularly, only when soil conditions are favorable for the microorganisms that change it into this form. Considered as a crop produced within the soil, these diminutive performers fit into the categories of soil fertility and duplicate the requirements demanded by good plant growth above the soil. Inasmuch as long time treatments modify microbial activities within the soil, so they will in turn influence the crop yields. Laboratory studies were undertaken to evaluate the nitrate accumulating capacities by the soils, which have undergone different long-continued cropping and soil treatments on Sanborn Field[2] as a measure of the influence by these on the microbial behavior and thus on the soil's capacity to deliver plant nutrients to the growing crop.

---

[1] W. A. Albrecht, "Nitrate Production in Soils as Influenced by Cropping and Soil Treatments," Research Bulletin 294 (Missouri Agricultural Experiment Station, October 1938).
[2] Sanborn Field was established in 1888 by J. W. Sanborn. The crop of 1938 was the fiftieth and treatments have been carried out as originated with some modifications.

Plan of the Study: Sanborn Field Plots Used

The treatments given the soils in the field during fifty years divide themselves into those influences arising from (a) the system of cropping, and (b) the soil treatments as fertility additions. Plots were selected for study of the former influence so as to include (a) continuous cropping to wheat, to oats, to corn and to timothy; (b) a rotation extending over three years with corn, wheat, and clover; (c) a rotation extending over four years with corn, oats, wheat, and clover, the last of which was replaced more recently by soybeans; and (d) a rotation over six years with corn, oats, wheat, clover, timothy, and timothy as the cropping order. For study of the influence by soil treatments as fertility additions, plot selections were made to permit comparisons of manure and no manure, superphosphate and no superphosphate, commercial fertilizer and no fertilizer, ammonium sulfate and sodium nitrate, and limestone and no limestone.

Plate 1. Sanborn Field, where fifty years of different soil treatments and cropping systems reflect their effects on the present soil fertility levels.

Laboratory Procedures

Since such studies involve numerous chemical determinations, only a few plots were studied each year from 1921 to 1937 inclusive. The samples were taken regularly at the same season and the studies carried out from about February 15 to May 15. The soils taken into the laboratory were not dried, but were used at optimum moisture in quantities equivalent to 100 grams dry soil. They were placed in jelly tumblers in sufficient numbers to permit the treatments in duplicate for chemical determinations at fortnightly intervals during ten or twelve weeks. The plots studied each year were chosen to represent comparable cases of field treatments, as manure and no manure, or fertilizer and no fertilizer with a single crop, or for a rotation. Laboratory conditions and methods of determination were standardized as far as possible so that the data of one year might be comparable to those of any other year.

The laboratory treatments consisted of (a) no treatment, (b) addition of one gram of powdered limestone (100-mesh), (c) addition of three grams of dry, finely chopped sweet clover, and (d) addition of these amounts of both the items per one hundred grams of soil. The sweet clover was used from the same supply during most of the tests. This was collected and dried indoors without leaching loss and contained 2.21 per cent of nitrogen on the air-dry basis. The soils in the tumblers were maintained at optimum moisture as far as possible by occasional weighings and moisture additions. They were incubated at room temperature. Though the fertility additions to the soils in the laboratory represent greater concentrations than are used in the field, they were the same throughout and serve, therefore, as comparable cases. For convenience, the laboratory treatments will be designated as (a) no treatment, (b) lime or limestone, (c) organic matter, and (d) lime and organic matter.

Sufficient tumblers of each soil were prepared to provide duplicates of each laboratory treatment for chemical study at the outset and every two weeks thereafter for ten weeks and sometimes for twelve weeks. For each fortnightly determination the soil in the tumbler was divided into two equal parts. One-half was used for determining the ammonia nitrogen. The other served for determination of the nitrate nitrogen. For the former, the moist soil was placed into a Kjeldahl flask; magnesium oxide, some paraffin and water were added; and the ammonia distilled over into the standard acid by means of a flow of air and steam. For the latter, or nitrate nitrogen determination, the soil was dried, the moisture loss determined, and the nitrates measured by the phenoldisulphonic acid method. In the first few studies, the nitrates were determined by extraction, reduction with De Varda's metal and distillation into standard acid. All calculations for both ammonia and nitrate nitrogen were then based on fifty grams of water-free soil.

Since the ammonia accumulation is the precursor of the nitrate production, the latter reflects the former. Consequently, the details regarding the ammonia contents are omitted in this study and attention will be given only to the accumulations of nitrate nitrogen.

Field Treatments of Soils Used

A total of 39 plot studies were made with eight nitrate determinations at each fortnightly period during twelve weeks. With more than two thousand such determinations, complete presentation of the tabulated data scarcely seems feasible. Among these 39 plot studies, only 28 individual plots were involved. Three plots were used three times, four were included twice, and 21 were used only once. The plot numbers, their cropping systems and soil treatments are assembled in Table 1. Plots 6 and 7 listed in the table are omitted in the reported data because of their irregularities in cropping history.

TABLE 1: Plots on Sanborn Field under long continued treatments used for study of nitrate accumulations.

Plot Number	Times Used	Cropping System	Soil Treatment
1	1	Rotation—six year	Manure 6 T. annually Rock Phosphate
2	2	Wheat—continuous	Fertilizers (a)
6	1	Cowpeas—continuous	...
7	1	Cowpeas—continuous	...
9	3	Wheat—continuous	...
10	2	Wheat—continuous	Manure 6 T. annually
12	1	Rotation—six years	Manure 3 T. (b)
13	1	Rotation—six years	...
14	1	Rotation—six years	Manure 3 T. Superphosphate
15	1	Oats—continuous	Manure 6 T.
16	1	Oats—continuous	...
17	2	Corn—continuous	...
18	1	Corn—continuous	Manure 6 T.
19	1	Rotation—six years	Manure 3 T.
20	1	Rotation—six years	Manure 6 T.
21	3	Wheat—continuous	Superphosphate
22	1	Timothy—continuous	Manure 6 T.
23	3	Timothy—continuous	...

24	1	Wheat—continuous	...
25	1	Rotation—three years	Manure 6 T.
26	1	Rotation—three years	Manure 3 T.
27	1	Rotation—three years	...
29	1	Wheat—continuous	Ammonium sulfate
30	1	Wheat—continuous	Sodium nitrate
34	1	Rotation—four years	Manure 6 T.
35	2	Rotation—four years	...
37	1	Rotation—four years	Fertilizer (c)
38	2	Rotation—four years	Fertilizer

(a) Applications were large enough to replace the fertility equivalent of nitrogen, phosphorus, and potassium of a forty-bushel crop.
(b) The three-ton applications are not made annually, but are the equivalent of three tons annually. The six-ton applications are all made annually.
(c) This is a mixed fertilizer applied with the wheat and the corn.
More complete history of these plots is presented in Missouri Agricultural Experiment Station Bulletin 182.

Results of the Study

Since the numerous determinations make the presentation of all the data impossible, consideration will be given to the nitrate accumulations according to (a) the influences by the cropping systems, and then (b) the influences by the soil treatments. These accumulations may be shown as related to the time period by means of graphs giving the nitrate nitrogen per fifty grams of soil at intervals of two weeks during ten weeks. They may be presented as nitrate increases in these same amounts of soil during a six-week period represented by the difference between the average of the second and fourth weeks and the average of the eighth and tenth weeks. In addition, the plots may be represented in their nitrate accumulation as measured by the mean of the six determinations for the entire period or ten weeks under observation. As a final figure, the mean of all determinations of all laboratory treatments may be used. Any or all of these measures may be used in presenting the data.

Influence of the Cropping Systems under Constant Soil Treatments

Continuous cropping—Four continuous crops, namely corn, oats, wheat, and timothy—all non-legumes—were represented. Each of these included two plots, one of which was given no soil treatment, and the other given manure at the rate of six tons annually. The two soil treatments are combined for the consideration of the influence by each cropping system. The data are presented together for these four different cropping systems on graphs in Figures 1, 2, 3, and 4. These represent the laboratory treatments of no treatment, limestone, organic matter, and lime and organic matter, respectively.

There was little difference in the nitrate accumulating capacity of the soil according to the crops grown on it, whether corn, oats, wheat, or timothy. In attempting to place these crops in an order of significant effects, they arrange themselves differently as the soils were given the different laboratory treatments. A study of the graphic presentation of the data in the figures for the soils taken from under these crops and given the different laboratory treatments shows the curves so closely intertwined that it is impossible to detect significantly superior influence by any one crop.

When the data are handled in other arrangements as given in Table 2, there are no more significant suggestions of relations to the crops. When the nitrate increase during a six week period is taken as the criterion of differences in nitrate production, no constant place in order can be assigned to each crop. If the mean of the six determinations made on any one laboratory treatment during the ten weeks is considered as the measure of the nitrate accumulating capacity, then the soil under timothy is consistently high, though oats surpass it slightly in two instances. When a mean of all the laboratory treatments for all determinations extending over ten weeks is taken as a measure of the crop influence on nitrate production, then the crops arrange themselves in the following order: timothy, wheat, oats and corn for lowered nitrate levels in their respective soils. The first three are of almost equal value, save that the oats are the lowest. One cannot say after the soil had been under each of these crops for a long period of time that it has been modified more or less in its capacity to produce nitrates from added materials because one or the other of these particular crops was grown on it. The low level for all of these continuous crops rather than a differentiation amongst them is the significant fact.

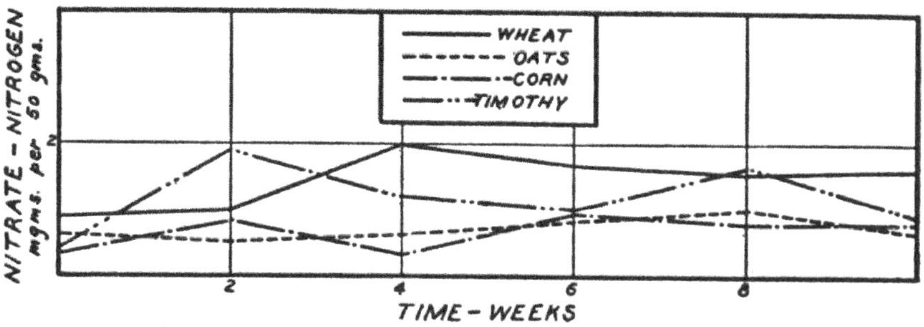

Figure 1. Nitrate accumulations in soils as influenced by long periods of continuous cropping when given No Treatment in the laboratory. (Soils from Sanborn Field)

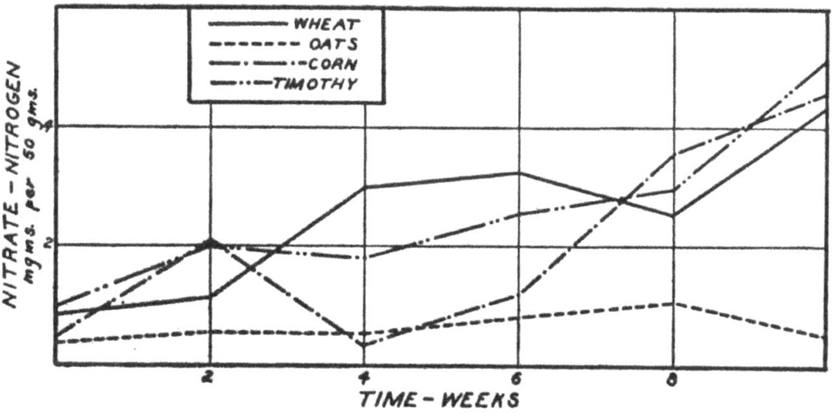

Figure 2. Nitrate accumulations in soils as influenced by long periods of continuous cropping when given Lime in the laboratory. (Soils from Sanborn Field)

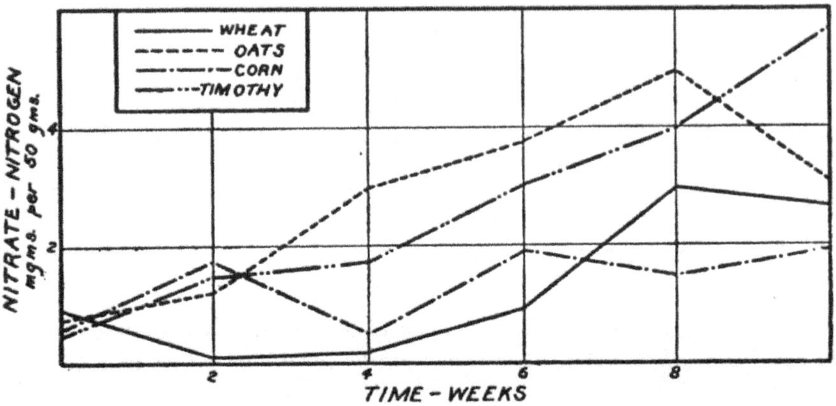

Figure 3. Nitrate accumulations in soils as influenced by long periods of continuous cropping when given Organic Matter in the laboratory. (Soils from Sanborn Field)

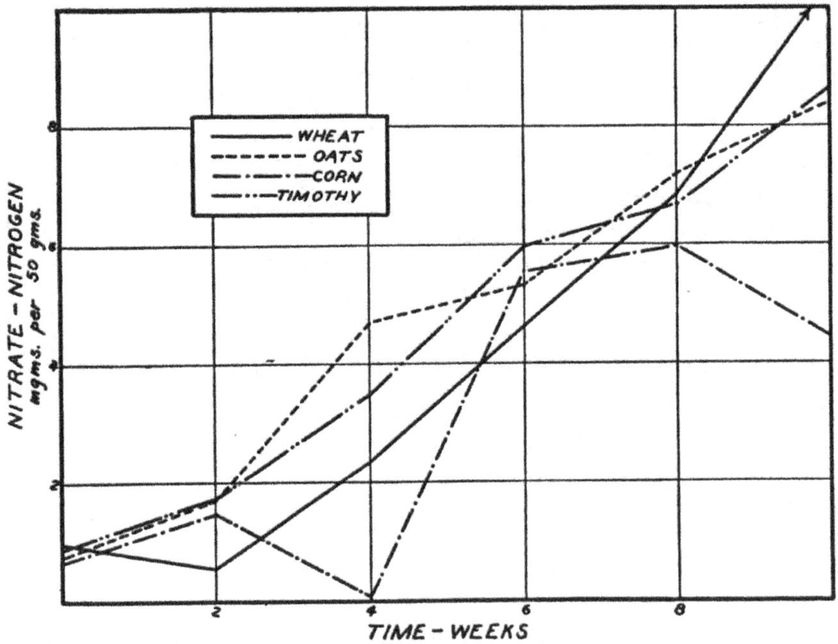

Figure 4. Nitrate accumulations in soils as influenced by long periods of continuous cropping when given Lime and Organic Matter in the laboratory. (Soils from Sanborn Field)

TABLE 2: Nitrate accumulation as influenced by different crops (Mgms. nitrate nitrogen per 50 gms. soil.)				
Crop	Laboratory Treatments			
	No Treatment	Limestone	Organic matter	Lime and organic matter
As increases during six weeks				
Corn	.23	2.74	.62	4.76
Oats	.48	.41	2.57	4.72
Wheat	.13	1.40	2.76	7.11
Timothy	.29	3.85	1.72	5.22
As mean of six determinations during ten weeks				
Corn	.64	2.08	1.39	3.06
Oats	.77	.66	3.04	4.75
Wheat	1.48	2.56	1.32	4.30
Timothy	1.21	2.62	2.76	4.59
As mean of all determinations of four treatments during ten weeks				
Corn—1.79	Oats—2.30	Wheat—2.41	Timothy—2.79	

Rotations—The rotations, unlike the continuous crops, differentiate themselves readily for their effects on nitrification according to the graphical data given in Figures 5, 6, 7, and 8. It is evident that the nitrates in the soil from the six-year rotation were not the equal of those in soils from the four-year and three-year rotations. They were higher in the three-year rotation than in the four-year rotation. Irrespective of which set of data as assembled in Table 3 is used in determining the order of influence by these rotations, they fall into the order of largest nitrate accumulation for short, or three-year rotation, and decreasing amounts for the lengthening of the rotation.

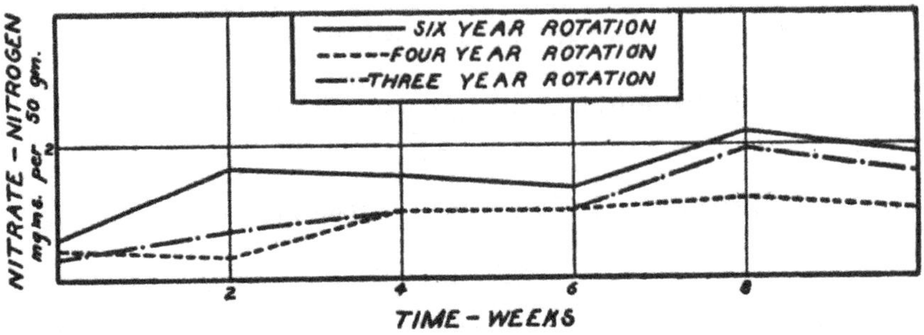

Figure 5. Nitrate accumulations in soils as influenced by long periods of rotation cropping when given No Treatment in the laboratory. (Soils from Sanborn Field)

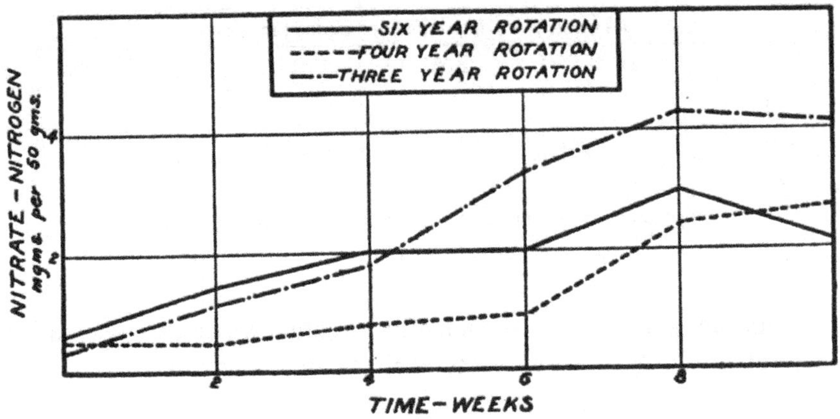

Figure 6. Nitrate accumulations in soils as influenced by long periods of rotation cropping when given Lime in the laboratory. (Soils from Sanborn Field)

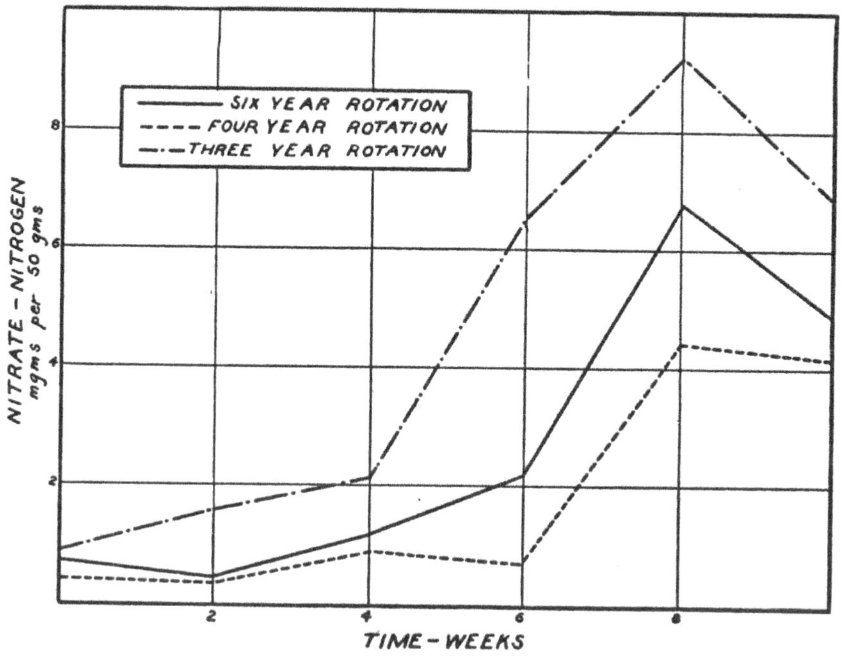

Figure 7. Nitrate accumulations in soils as influenced by long periods of rotation cropping when given Organic Matter in the laboratory. (Soils from Sanborn Field)

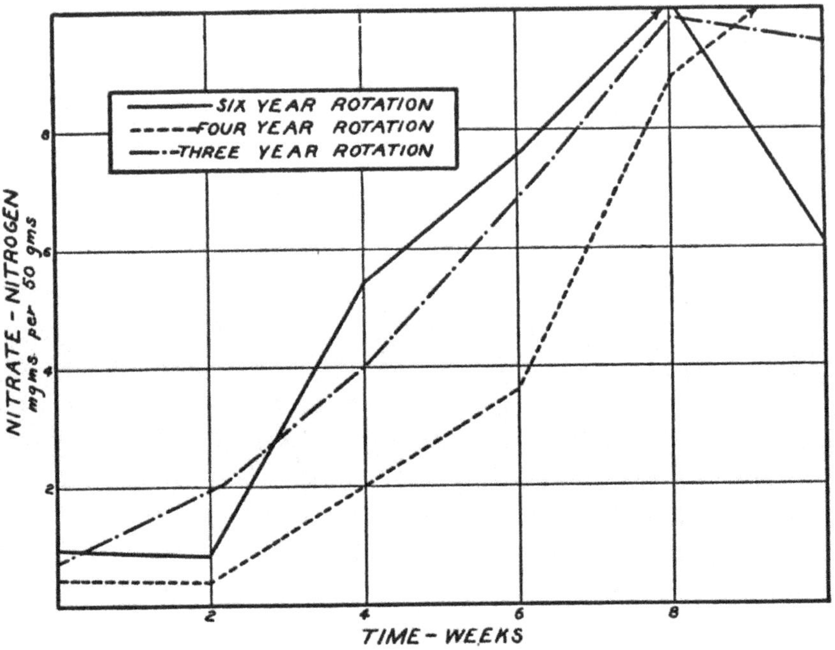

Figure 8. Nitrate accumulations in soils as influenced by long periods of rotation cropping when given Lime and Organic Matter in the laboratory. (Soils from Sanborn Field)

TABLE 3: Nitrate accumulations as influenced by different rotations (Mgms. nitrate nitrogen per 50 gms. soil.)

Rotations	Laboratory Treatments			
	No Treatment	Limestone	Organic Matter	Lime and organic matter
As increases during six weeks				
Six year rotation	.49	.56	3.69	7.72
Four year rotation	.33	1.01	5.46	5.93
Three year rotation	.86	2.73	6.13	1.76
As mean of six determinations during ten weeks				
Six year rotation	.85	1.30	1.86	4.42
Four year rotation	1.40	1.89	2.73	5.22
Three year rotation	1.11	2.43	4.55	5.48
As mean of all determinations of four treatments during ten weeks				
Six year rotation—2.11.	Four year rotation—2.81.	Three year rotation—3.39.		

The attachment of significance to the data showing an influence by the rotation on nitrate accumulation may seem doubtful when the separate crops in continuation were of less influence. However, the continuous crops were non-legumes while each of the rotations included a legume for one year. The influence by the rotation was greater as the period of the rotation was shorter, or as this one legume recurred more often in the involved period of the soil's history. Though the clover was not an excellent crop on the unmanured plots in the later period of the cropping history, it was generally of good growth on those given the field treatment of manure. Since the manured and unmanured soils were combined in these data, the greater number of clover occurrences in the cropping history may be a partial reason for the higher nitrate accumulation in the shorter rotation. If these are the facts, then the growing of the clover crop—quite different from the crops of corn, oats, wheat and timothy—is instrumental in giving to the soil on which it grows a higher capacity to transform added materials into nitrate accumulations.

These data suggest then that the non-legume crops can have little influence, per se, on the nitrification activity by the soil, but that a legume crop, possibly through its contribution of organic matter of a narrow carbon-nitrogen ratio, or because of some other conditions, encourages greater nitrate production in the soil.

Influence of Soil Treatments under Constant Conditions of Cropping

Soil treatments in the field, much more than cropping systems, had a significant influence on the soil's activity as a nitrate producer when brought into the laboratory. In case of the field treatments using manure and no manure, both continuous cropping and rotation systems were included. The results from fourteen plot studies were combined under no manure, and from nine plots under manure. Six year rotations and continuous wheat as four plots were combined under the phosphate to be compared with the four corresponding no phosphate treatments. These same four combined with those given mixed fertilizers and with those given nitrogen only, were taken as a group of nine plots as the fertilizer treatment. They were contrasted with the same cropping treatments given no fertilizer. One plot with ammonium sulfate and another with sodium nitrate, both in continuous wheat, and a pair of plots in a four year rotation without and with lime respectively, completed the comparisons of nitrate accumulations as influenced by soil treatments.

Manure—The data for the accumulation of nitrate during the ten week period as illustrated in Figure 9 show a decided influence by the manure additions. It is evident from these curves that the soils which had been receiving manure these many years were consistently higher in nitrate content than those not obtaining it. When their abilities to increase their nitrates in a six week period, or their mean nitrate contents over ten weeks, are calculated as given in Table 4, the significance of the manure as a nitrate increaser is obvious.

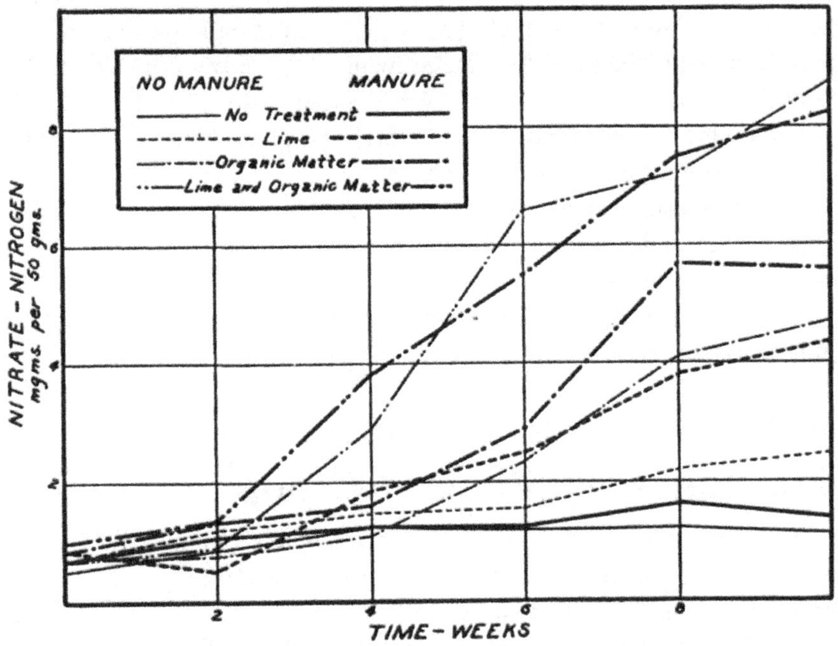

Figure 9. Nitrate accumulations in soils as influenced by long periods of Manure and No Manure additions when given different treatments in the laboratory. (Soils from Sanborn Field)

TABLE 4: Nitrate accumulations as influenced by manure additions over long periods.				
Field Treatments	Laboratory Treatments			
	No Treatment	Limestone	Organic Matter	Lime and organic matter
As increases during six weeks				
No manure	.21	.85	3.26	5.31
Manure	.47	2.34	4.00	5.69
Increase by manure	.26	1.49	.74	.38
As mean of six determinations during ten weeks				
No manure	1.02	1.54	2.32	4.49
Manure	1.24	2.33	3.01	4.91
Increase by manure	.22	.79	.69	.42
As mean of all determinations of four treatments during ten weeks				
No manure—2.34	Manure—2.87	Increase—22 per cent		

It is interesting to note the influence of the lime addition in the laboratory on these soils manured and unmanured for so many years. After lime was on the soil for six weeks in the laboratory, the unmanured had increased its nitrate nitrogen by .85 mgms, or an increase of .64 mgms. greater than without added lime. In case of the manured soil given lime similarly, the increase of accumulated nitrate nitrogen was 2.34 mgms. or 1.87 mgms. greater than without lime additions in the laboratory. This indicates that, during the long continued manuring, this soil accumulated a store of nitrogenous organic matter the conversion of which to nitrates did not take place because of the need for liming. Lime addition now hastened the conversion of this residual matter into the soluble form.

The data of Table 4 point further to the significance of the manure. During six weeks in the laboratory the increase was twice as great in the check, manured soil as in the unmanured. This difference amounted to .26 mgms., or the equivalent of ten pounds of available nitrogen per acre. The increases in nitrate nitrogen content for the three different treatments in the laboratory during the six weeks show the manured soil higher by more than this figure. This same holds true for the mean during ten weeks. When all determinations for all treatments are averaged, the manured soil is 22 per cent higher than the unmanured. This points out that the manured soil responds more to lime, and that it converts its organic matter additions more effectively. It indicates too that as the soil is depleted, the restorative treatments are relatively less effective, or, as the fertility level drops, the difficulty in raising it becomes relatively greater.

Superphosphate—As a means of studying the influence of superphosphate field treatments on the soil's capacity to accumulate nitrates, four plots including the six year rotation and the continuous wheat were studied. The data for the accumulation of nitrate are assembled as graphs in Figure 10. They are assembled as averages for numerical evaluation in Table 5.

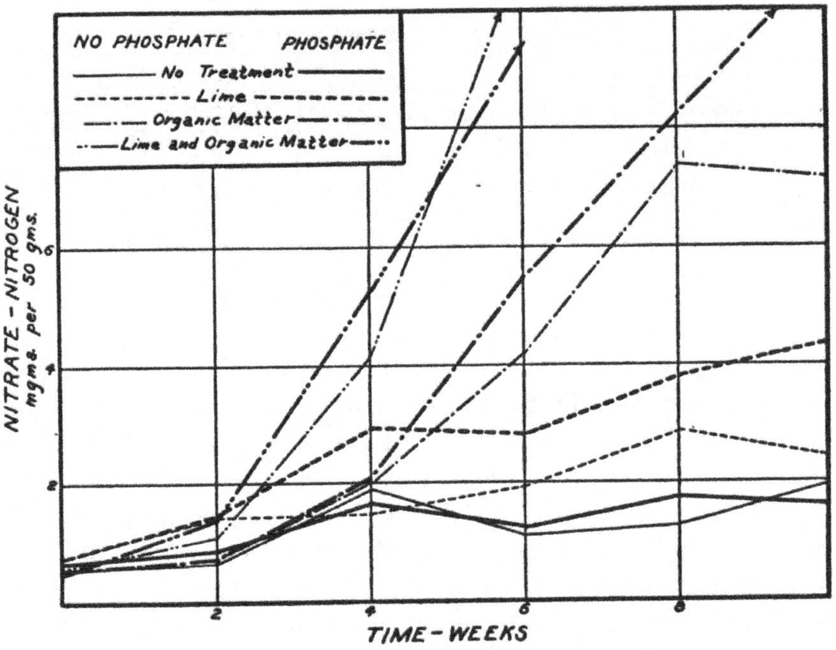

Figure 10. Nitrate accumulation in soils as influenced by long periods of Superphosphate and No Superphosphate additions when given different treatments in the laboratory. (Soils from Sanborn Field)

The influence of the phosphate is readily evident. It is interesting to note that the no treatments are not different because one had phosphate. They suggest that, whether given phosphate or not, the nitrate is at a low and relatively constant level. This is different from the effects by manure which contributed to a higher nitrate for the soil given no treatment in the laboratory. When these soils were limestone only, that which already had received phosphate in the field was more effective in liberating nitrogen from its reserve than the soil without phosphate. When organic matter was added, the soil with phosphate made more effective use of this nitrogen addition to be converted into nitrates. When both lime and organic matter were added, their behaviors were similar to those for the additions of organic matter only. Hence this suggests that phosphating the soil is similar to a liming treatment. The general superiority of the phosphated soil as a nitrate producer is shown by the mean of all determination in the Table 5 where the difference amounts to 19 per cent.

TABLE 5: Nitrate accumulations as influenced by phosphate additions over long periods. (Mgms. nitrate nitrogen per 50 gms. soil.)

Field Treatments	Laboratory Treatments			
	No Treatment	Limestone	Organic Matter	Lime and organic matter
As increases during six weeks				
No phosphate	.31	1.27	5.96	9.12
Phosphate	.30	1.92	8.08	9.70
Increase by phosphate	.01	.67	2.12	.58
As mean of six determinations during ten weeks				
No phosphate	1.24	1.78	3.66	6.73
Phosphate	1.25	2.65	4.62	7.06
Increase by phosphate	.01	.87	.96	.33
As mean of all determinations of four treatments during ten weeks				
No phosphate—3.35	Phosphate—3.89	Increase—19 per cent		

That phosphate has an effect similar to liming is indicated by the response when organic matter only was added as compared to that when both lime and organic matter were supplied. According to Table 5, under "organic matter," the gain in nitrates in six weeks was 8.08 mgms. on plots receiving phosphate, but only 5.96 by soils with no phosphate. The phosphated soil delivered 2.12 mgms. more. When lime and organic matter were added to the soil with no phosphate, the nitrate increase was 9.12 mgms. or an amount greater by 1.04 mgms. as a consequence of liming. The phosphate treated soil brought with it a capacity to give nitrate increase in 6 weeks from added organic matter that was 2/3 as great as the untreated soil acquired when given both lime and organic matter. This influence on nitrate accumulation by phosphate that tended to duplicate the influence by liming was noticeable regularly in these studies. If the effects by phosphate are similar to those by liming, this suggests that calcium, their common element, which makes up a part of the superphosphate (approaching 25 per cent) may be the responsible item. If these are the facts, we may not have been ascribing the value of superphosphate to the wholly correct item. We may, then also, expect less value from the more concentrated phosphates which carry little or no calcium.

Fertilizers—When the data for the plots with superphosphate were combined with those from a continuous wheat plot given heavy fertilizer treatment, with the same crop given ammonium sulfate, one given sodium nitrate, and a four year rotation using fertilizers, the results from these nine plots are closely similar to those without these treatments as shown in Figure 11. The curves are so closely similar in the no treatments for both the no fertilizers and for the fertilizers as to be almost inseparable. The small differences between the two soils are confirmed by the data of Table 6. The values are closely similar except under organic matter. Even these are lesser differences than those shown by the phosphate treatment included herewith and the influence of which is reflected in this larger group.

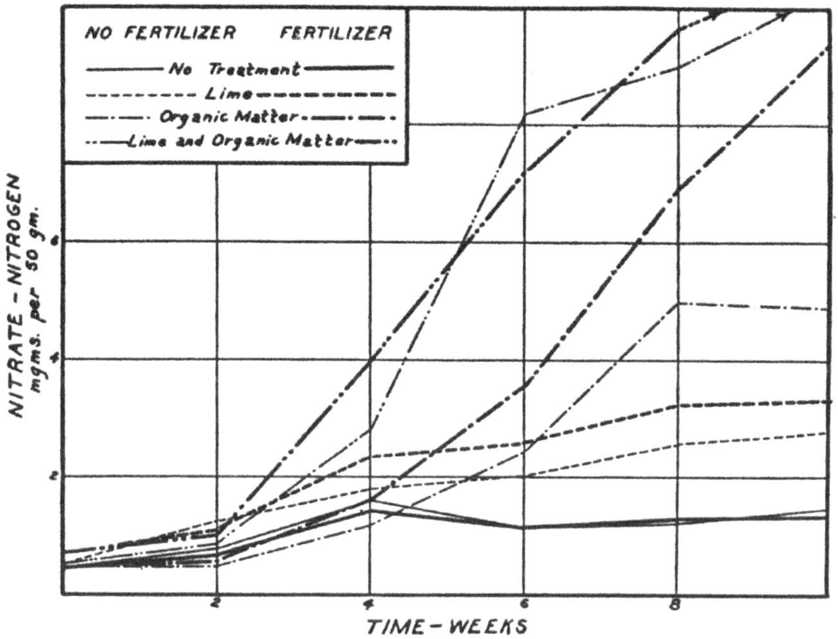

Figure 11. Nitrate accumulations in soils as influenced by long periods of Fertilizer and No Fertilizer additions when given different treatments in the laboratory. (Soils from Sanborn Field)

TABLE 6: Nitrate accumulations as influenced by fertilizer additions over long periods.

Field Treatments	Laboratory Treatments			
	No Treatment	Limestone	Organic Matter	Lime and organic matter
As increases during six weeks				
No fertilizer	.15	1.13	4.02	7.45
Fertilizers	.30	1.48	6.97	7.63
Increase by fertilizer	.15	.35	2.95	.18
As mean of six determinations during ten weeks				
No fertilizer	1.09	1.80	2.40	5.38
Fertilizer	1.09	2.35	3.74	5.56
Increase by fertilizer	.00	.55	1.34	.18
As mean of all determinations of four treatments during ten weeks				
No fertilizer—2.67	Fertilizer—3.18	Increase—19 per cent		

Ammonium sulfate and sodium nitrate—The data for the nitrate accumulation by the soils under continuous wheat given ammonium sulfate and sodium nitrate treatments are assembled in Figure 12. It is evident that the untreated soils behave similarly, but the effect by lime is the greater on the ammonium sulfate plot. This would be expected from our knowledge of the base depleting activity, or the propensity toward acidity, of a soil treated continuously with ammonium sulfate. Organic matter addition alone was more effective on the soil with sodium nitrate. When both limestone and organic matter were added, the sodium nitrate soil was slightly more effective in producing nitrates.

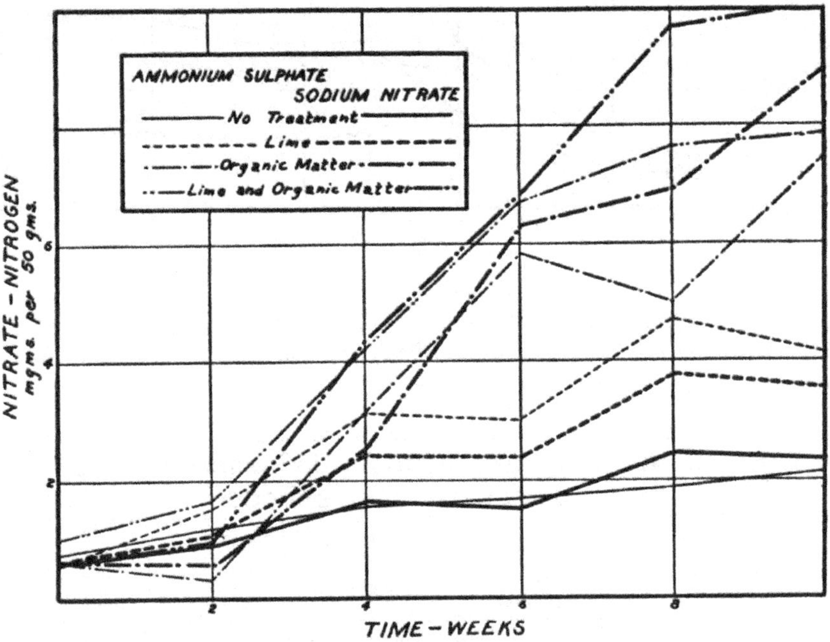

Figure 12. Nitrate accumulations in soils as influenced by long periods of Ammonium Sulfate and Sodium Nitrate additions when given different treatments in the laboratory. (Soils from Sanborn Field)

Limestone—Only two plots were available that were comparable for measuring the influence of limestone over a long period on the nitrate accumulating activity. The data for these are given in Figure 13. The interesting feature is the pronounced effect by laboratory additions of limestone on the unlimed soil, and the much smaller effect by it on the soil from the plot that was limed in the field. Further, the limed soil converted the added organic matter into nitrates at a much higher rate than the unlimed soil. The limed soil was earlier in its effects when both lime and organic matter were added. It gave decided increase at two weeks while the unlimed gave no response until after four weeks. This earlier nitrate production as a result of liming may be a significant item in crop production, and may be an effect not commonly appreciated.

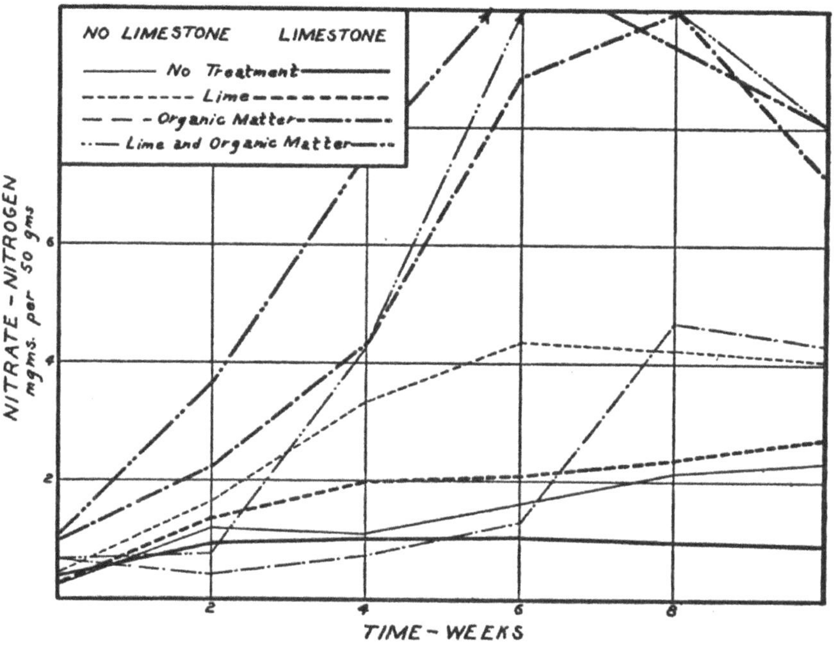

Figure 13. Nitrate accumulations in soils as influenced by long periods of Limestone and no Limestone additions when given different treatments in the laboratory. (Soils from Sanborn Field)

Summary

In order to illustrate more clearly the influence by the continuous cropping history on the soil's capacity to accumulate nitrates at only low figures, as compared to the distinct influence by the rotations and the soil treatments with their higher figures, the average nitrate content during the ten weeks may be taken as a fair criterion for each treatment. Using this figure in terms of pounds per acre, the results of the many foregoing tables are assembled in more generalized form as a summary in Table 7.

TABLE 7: Nitrate nitrogen levels in soils under different laboratory treatments as influenced by the past history of cropping and soil treatments. (Soils from Sanborn Field)

(Pounds Nitrogen per Acre)

Cropping History and Field Treatment		Laboratory Treatments				Mean of all Treat-ments	Increase over lowest item
		No Treatment	Limestone	Organic Matter	Lime and Organic Matter		
Continuous Crops	Corn	25	83	55	122	71	...
	Oats	30	26	121	190	92	21
	Wheat	59	102	52	172	96	25
	Timothy	48	104	110	183	111	40
Rotations	Six year	34	52	74	176	84	...
	Four year	56	75	109	208	112	28
	Three year	44	97	182	219	135	51
Soil Treatments	None	40	61	92	179	93	...
	Manure	49	93	120	196	114	21
	Phosphate	60	106	184	282	155	62
	Fertilizers	43	94	149	222	127	34
	Ammonium Sulfate	60	113	150	193	129	36
	Sodium Nitrate	62	87	174	216	134	41
	Lime	36	74	223	267	150	57

If the general mean of all determinations is considered as a basis for comparison, then the soil under timothy had a higher nitrate accumulating capacity than that under wheat, or under oats—which two were in close agreement—and these had a higher nitrate producing power than the soil from corn. The nitrate accumulating capacity by the soils in different rotations takes the following order very clearly for decreasing levels; three year, four year, and six year rotations.

As for the influence by soil treatments, the largest nitrate level was supplied by soils with phosphates. These were followed closely by the limed soil. Both of these were higher than any other treatment. If a share of the influence by phosphorus was due to the calcium it carried, then the liming receives double emphasis as an important factor in nitrification. Even if this extra emphasis is not separated out to be given to calcium, joint importance can be given to limestone and phosphate treatments on these older experimental plots. The use of nitrogenous fertilizers, mixed fertilizers, manure, and the three year rotation—with their additions of nitrogen—all gave these soils greater capacity to deliver nitrate nitrogen over the time period of the test. This gives emphasis to the need for supplying some nitrogenous materials to these soils if they are to deliver soluble nitrogen at a high rate during the growing season. The need by these soils for mineral fertilizers, particularly limestone and phosphate, if they are to give improved nitrate delivery is the outstanding suggestion from these studies.

Conclusions

The studies on nitrate accumulating capacities of soils under long continued treatments on Sanborn Field point out that the particular cropping system on the soil was of lesser importance than soil treatments in determining the ability of the soil to accumulate nitrates. Soil from corn continuously had the lowest nitrate producing capacity. Soils from continuous wheat and from oats were of equal capacity, but slightly higher than that from corn. Soil from timothy was above that from these two cereal grains.

Soils from under crop rotations had greater capacities for giving nitrates as the rotations were shorter. Since clover occurred only once in each of three, four, or six year rotations, this suggests that this improved nitrate delivery in the shorter rotation may be due to the more numerous crops of clover to make their greater additions of nitrogenous matter to the soil.

Soil treatments were of significance in giving improved nitrate accumulation. The use of phosphates and limestone was most pronounced in effects on nitrification, even though they added no nitrogen to the soil. Their addition would then hasten depletion of the soil nitrogen through its conversion into the soluble form. The addition of nitrogenous fertilizers, mixed fertilizers, and manure, were all effective in improving nitrate accumulation in the laboratory studies. Soils which had been given these nitrogenous materials as field treatments had a better capacity to deliver soluble nitrogen when under test.

These facts point forcibly to the need for supplying the soil with nitrogen to be nitrified, and then also with the minerals as essentials in the functioning of the microorganisms that convert the nitrogen into the available form. These results emphasize not only the need by the soils for nitrogenous materials to be nitrified, but also suggest the need for more consideration of the importance of the mineral fertilizers as they serve in promoting better microbiological soil condition for the production of soluble nitrogen in the soil. They point out that we need to consider the soil as a supply of plant nutrients, and, in addition, as the factory operated by microbiological activity to convert these into forms and supplies sufficient for the growing plants. Long continued treatments have had their influence on the soil in both these respects.

Index

Other Titles from Acres U.S.A. Books:

Restoration Agriculture
By Mark Shepard

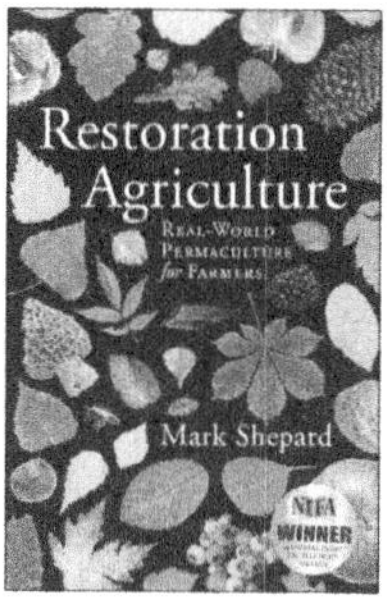

Around the globe, most people get their calories from "annual" agriculture—plants that grow fast for one season, produce lots of seeds, then die. Every single human society that has relied on annual crops for staple foods has collapsed. *Restoration Agriculture* explains how we can have all of the benefits of natural, perennial ecosystems and create agricultural systems that imitate nature in form and function while still providing for our food, building, fuel and many other needs—in your own backyard, farm, or ranch. Based on real-world practices, this book presents an alternative to the agriculture system of eradication and offers exciting hope for our future.

Hands-On Agronomy
By Neal Kinsey & Charles Walters

The soil is more than just a substrate that anchors crops in place. An ecologically balanced soil system is essential for maintaining healthy crops. *Hands-On Agronomy* is a comprehensive manual on effective soil fertility management providing many on-farm examples to illustrate the various principles and how to use them. The function of micronutrients, earthworms, soil drainage, tilth, soil structure, and organic matter is explained in thorough detail.

Neal Kinsey demonstrates that working with the soil produces healthier crops with a higher yield. To that end, he provides an understanding of eco-agriculture as a viable enterprise that is both naturally and commercially sustainable. This work offers advice that is both highly valuable and applicable to the farmer of any level and promotes an ecological understanding of the farm from the ground up.

Water for Any Farm
By Mark Shepard

Written as a companion to the bestseller, *Restoration Agriculture*, this book will help farmers capture water in areas they want to, and avoid having water flow immediately to the low point. The result? Less water expense, healthier crops and livestock, and less erosion ... just to name a few.

What you will read in this book is a distillation of over 25 years of on-the-ground experience working with and modifying the Yeomans' Keyline Plan. From the back yard suburbs to 10,000-acre ranches and everywhere in between, from permafrost mountainsides just shy of the Arctic Circle, to equatorial boulder fields of East Africa, areas with 300 inches of rain per year to those with less than 3 inches, I have personally installed systems based on the Keyline design methodology and its modified forms.

bookstore.acresusa.com

Visit acresusa.com
to learn about our magazine, events, books, and other educational opportunities.

Made in the USA
Monee, IL
13 October 2026

65495369R00164